Higher National Computing Tutor Resource Pack

Higher National Computing Tutor Resource Pack

Second edition

Howard Anderson

Sharon Yull

Bruce Hellingsworth

 Routledge
Taylor & Francis Group

LONDON AND NEW YORK

First published by Newnes
This edition published 2011 by Routledge
2 Park Square, Milton Park, Abingdon, Oxon OX14 4RN

Simultaneously published in the USA and Canada by Taylor & Francis Group,
711 Third Avenue, New York, NY 10017, USA

Routledge is an imprint of the Taylor & Francis Group, an informa business

First published 2003

Second edition 2004

British Library Cataloguing in Publication Data
A catalogue record for this book is available from the British Library.

ISBN 0 7506 61267

Contents

Introduction		vii
Explanation		viii
Internet, the main search engines		viii
Effective searching of the Internet		x
Worksheet		xi

1   Enrichment   **1-01**
   1.1   Moore's law: is it true?   1-01
   1.2   Enrichment material   1-02
   1.3   Enrichment material   1-07
   1.4   ASCII and Unicode   1-11
   1.5   Enrichment material   1-13
   1.6   Assignment   1-48

2   Systems analysis   **2-51**
   Rationale   2-51
   2.1   Systems analysis lifecycle   2-51
   2.2   Systems analysis tools and techniques   2-54
   2.3   Systems investigation   2-81
   2.4   Functional and data modelling   2-84

3   Programming concepts   **3-109**
   Worksheet Units 3.1 to 3.8 Pascal   3-109
   Answers to worksheets   3-117

4   Concepts of database design   **4-129**
   4.1   Activity One – File-based systems   4-129
   4.2   Activity Two – Data models   4-130
   4.3   Activity Three – Database Management Systems (DBMS) 4-130
   4.4   Activity Four – Database security   4-130
   4.5   Activity Five – Database design and editing   4-130
   4.6   Activity Six – Case study: The Organic Food Company 4-131
   4.7   Activity Seven – Database application cycle   4-136
   4.8   Activity Eight – SQL exercises   4-136
   4.9   Activity Nine – Data mining   4-137

5   Network assignment   **5-139**
   5.1   What you are required to do   5-139
   5.2   What you must deliver   5-139
   5.3   Possible solutions   5-139

6   Personal skills development   **6-143**
   6.1   Activity One – Planning and review   6-143
   6.2   Activity Two – Report writing skills   6-144
   6.3   Activity Three – SWOT analysis   6-144
   6.4   Activity Four – Team roles   6-145

6.5	Activity Five – Verbal communication	6-146
6.6	Activity Six – Creating a CV	6-147
6.7	Activity Seven – Impact of verbal communication	6-148
6.8	Activity Eight – Producing written documents	6-148
6.9	Activity Nine – Using graphs and charts	6-153
6.10	Activity Ten – Presentations	6-153
6.11	Activity Eleven – Formal presentations	6-155
6.12	Activity Twelve – Freestyle presentation	6-155

7 Quality systems   **7-157**

Rationale   7-157

7.1	Quality assurance	7-157
7.2	Quality control	7-168
7.3	Project management	7-175
7.4	Systems development review	7-188

Appendix A Fact sheet   AA-219

Appendix B Fact sheet   AB-221

Introduction

This tutor's resource pack has been designed to provide activities for students to *do* rather than simply supply provide material extra to the book.

One of the main difficulties in preparing the book and this pack is in the setting of the correct level. Edexcel requirements are a good guide but often examples are required to add clarity. Computing is now such a large subject that no treatment at level 4 can possibly hope to cover every aspect; some aspects will necessarily be given more prominence than others. This pack will concentrate more on the items that require practice such as programming and communications, and less on sets of facts to be remembered.

It has been recognized that some aspects of the unit requirements are catered for using centre-specific equipment and software. For example, some centres may use Apple computers, others use PCs with linux, others still use PCs running any of the several versions of the Microsoft Windows operating system. For this reason, areas that are likely to be centre specific will not be found in this pack. In particular, practical exercises to cover the operating systems and networking parts of the units will have to be centre specific. Wherever possible, the tasks in this pack are generic in nature except the Pascal answers. These were prepared using Free Pascal, which runs on PCs.

The answers to each worksheet are examples only. Students should be encouraged to seek their own style so the answers in this pack should be used as a guide only. Programs and other testable items have only been tested to a standard appropriate to the course, i.e. not with the same rigour that would be applied to commercial software.

For 'fact finding' aspects of the units, students should be encouraged to use the Internet to the full. As computing moves at a great rate, only the Internet can give the latest information. Not so long ago the fastest PC available had a processor fitted that ran at 800 MHz, just a few months later, speeds of 1.2 GHz were available and at a lower cost than the 800 MHz machines. As this text is being prepared, machines of 3 GHz are commonplace and far faster machines will be available soon.

For this reason, one of the most useful skills that students should develop is use of the Internet. It is not often realized that successful use of search engines etc. depends more on the user's knowledge of the subject and their ability in the use of language than plain 'computer skills'. A historian is more likely to find good historical information than a 'computer whizz'.

Howard Anderson
Sharon Yull
Bruce Hellingsworth

 # Explanation

Internet, the main search engines

Some search engines have information (called a directory) organized by humans, others have information organized by computers. The second kind uses software (called a spider, robot or crawler) to look at each page on a website, extract the information and build an index. It is this index you search when using the search engine. Some people make a clear distinction between a directory and a search engine. Currently, the situation is not clear cut as many 'search engines' in fact use both methods.

The performance of a search engine depends critically on how well these indexes are built. It is also very important to remember that the whole business of search engines is in a state of constant change. Companies buy each other, change systems and indexes etc. Some rely on other people's information. There is no such thing as a static search engine!

Below are listed some of the main search engines.

AllTheWeb.com (FAST Search) http://www.alltheweb.com	One of the largest indexes of the web.
AltaVista http://www.altavista.com	One of the oldest crawler-based search engines on the web, it also has a large index of web pages and a wide range of searching commands.
AOL Search http://search.aol.com/	Uses the index from Open Directory and Inktomi and offers a different service to members and non-members.
Ask Jeeves http://www.askjeeves.com	Ask Jeeves is a human-powered search engine that introduced the idea of plain language search strings.
Direct Hit http://www.directhit.com	Direct Hit uses its own 'popularity engine' that depends on how many times a site is viewed to judge its ranking. This idea is not always successful as the less popular sites do not get a chance to rise, so popular ones remain popular. Direct Hit is owned by Ask Jeeves.
Google http://www.google.com	Google makes use of 'link analysis' as a way to rank pages. The more links to and from a page, the higher the ranking. They also provide search results to other search engines such as Yahoo.
HotBot http://www.hotbot.com	Much of the time, HotBot's results come from Direct Hit but other results come from Inktomi. HotBot is owned by Lycos.
Inktomi http://www.inktomi.com	You cannot query the Inktomi index itself, it is only available through Inktomi's partners. Some 'search engines' simply relay what is found in the Inktomi index.
LookSmart http://www.looksmart.com	LookSmart is a human-compiled directory of websites but when a search fails, further results are provided by Inktomi.
Lycos http://www.lycos.com	Lycos uses a human developed directory similar to Yahoo and its main results come from AllTheWeb.com and Open Directory.
MSN Search http://search.msn.com	Microsoft's MSN is powered by LookSmart with other results from Inktomi and Direct Hit.

Netscape Search
http://search.netscape.com

Netscape Search's results are from Open Directory and Netscape's own index. Other results are Google.

Open Directory
http://dmoz.org/

Open Directory uses an index built by volunteers. It is owned by Netscape (who are owned by AOL).

Teoma
http://www.teoma.com

A new search engine, launched in April 2002, that claims to be better than Google.

Yahoo!
http://www.yahoo.com

Yahoo! is a human-compiled index but uses information from Google.

Effective searching of the Internet

Some subjects are hard to find as the words used in a search engine often lead to many different subjects.

The task here is to find specific information and to eliminate all the non-relevant information. There is no 'answer' as such, you either find what you want or not.

Example: the ingredients of chocolate powder

If you want to know what is in chocolate powder, you might use the search string 'chocolate ingredients'.

Using this string, Google gave 417 000 'hits' and Lycos gave 7977. None of the hits at the top of the Google list were useful as they only contained the words 'chocolate' and 'ingredients'. One hit referred to a Parliamentary debate! Lycos gave fewer hits as the index is human organized so 'chocolate ingredients' is looked upon as a *subject*, not just *keywords*.

The key is to use whatever search engine/directory best suits your purpose. If you are searching for a 'subject', use a *directory*, if you are searching for specific information that can be isolated with a keyword or two, use a *search engine*. If necessary, use both.

To narrow the results, you can use the 'advanced' searches. There is nothing advanced about them!

In Google Advanced, these words were entered:

with all the words	chocolate
with at least one of the words	ingredient ingredients components
	substance analysis
without the words	cake biscuit cookie sponge

This simply results in the search string

chocolate ingredient OR ingredients OR components OR substance OR analysis − cake − biscuit − cookie − sponge

where the OR operator is self-explanatory and the ' − ' operator means leave out pages with this word. (You can type such a search string directly into most search engines. Their rules vary a little but most use + to include a word and − to exclude it.)

This search string gave a link that contained this information:

Sugar, modified palm kernel oil, modified milk ingredients, soya, lecithin, coco powder, artificial colour, vanilla and vanillin.

This is useful as it gives more keywords that you may not have thought of that should be used to narrow the search. Using these new keywords yielded a link that contained these ingredients:

cocoa processed with alkali, reduced minerals whey, maltodextrin, soy lecithin, tricalcium phosphate, salt, aspartame (non-nutritive sweetener), acesulfame potassium (non-nutritive sweetener), artificial and natural flavors.

The key point to note is that search engines or directories are simply an aid to searching, you must work at it to get the best results.

Can you trust the answers you find?

That depends on who published the information. The problem is no different when considering the Internet, magazines or books.

If the site owner is a private individual, the information may be correct but it is not likely to have been checked with great rigour. If it tallies with similar sites, the chances are good that it is correct but on the other hand, who copied who?

If it is an official site, the information will generally be an accurate reflection of the opinion of the site owners. Chocolate manufacturers will tell you chocolate is good for you, governments will tell you they serve you to the highest standards. Make your own judgement.

Academic sites often contain information from a more 'free thinking' group of people so will show a wide range of opinion. Specific scientific, technological or historical information is usually accurate, political views may be very varied. Again, make your own judgement.

Worksheet

Searching the Internet
Find out the following specific information. You are expected to use preliminary results to learn more about the subject, then use this new knowledge to search in more detail. You are not expected to know anything about the subject matter when you start.

Example
What is the name of the person who designed the world's first effective computer, which was used to decode German military messages during the Second World War.

First results yielded the name of the computer, *Colossus*. It was made to decrypt the German Lorenz (not Enigma) codes.

A new search using the word Colossus yielded the information *Tommy Flowers*. A genius. See http://www.codesandciphers.org.uk/

Your tasks
Find out the following specific information. Only give results that you would trust, i.e. if you find information that seems to tell you what you want but are not *sure* it is correct, ignore it.

1. What is the fastest current Intel Pentium microprocessor?
2. What is the maximum data capacity of a commercially produced DVD?
3. What is the fastest current microprocessor fitted in Apple G4 machines?
4. What are the main side effects of taking the drug aspirin over a long period?
5. How many people, soldiers and civilians from all countries were killed in the Second World War? Especially in this case, how can you be sure the answer you get is correct?
6. A British Prime Minister is quoted as having said 'There are lies, dammed lies and statistics'. Who was it?
7. It is said that a British Prime Minister once included in a speech the phrase 'Blood, sweat and tears'. Is it true or did he say something else?
8. Who wrote the famous poem 'If'?
9. How many standard floating point number formats are specified by the IEE?
10. Imagine you have just been loaned a 50 foot motor boat by a rich friend. It is moored at Southampton. Is it legal for you to take it alone across the channel to France?

1 Enrichment

1.1 Moore's law: is it true?

Moore's law

In 1965, Gordon Moore was the Research Director of the electronics company Fairchild Semiconductor. Some 3 years later he was to become one of the founders of Intel. He made an interesting prediction based on what had happened up to that time with memory chips (not microprocessors). He noticed that memory capacity doubled in capacity every 18–24 months. He then predicted this would continue, leading to an exponential growth in size and hence computing power. Market trends have shown it to have come true up until now with memory and with microprocessors; there are predictions that it will fail in the future but these predictions have themselves failed in the past as they have been made many times.

It seems that Dr Moore thinks it is still true. Notice, however, that in some people's eyes the 'law' has changed. It originally referred to the density of transistors on a piece of silicon. This is closely related to RAM capacity but many people take it to mean 'processor speed'. Transistor density is only one of the many factors that contribute to processor speed. You should also notice that Dr Moore refers to Intel products, not surprising considering he is a co-founder of the company, but as the marketplace is very competitive and other companies' products are comparable, it is reasonable to extrapolate his results to other firms.

On 10 May 2000, Gordon Moore gave an interview reviewing 'his' law. You can find this at http://www.usnews.com/usnews/transcripts/moore.htm.

There must be a limit to this improvement. This limit is often called a 'wall' and has been predicted for some time. At any one time, the technical difficulties that must be overcome to make ever better semiconductors are very severe. It is not always obvious that they will be overcome. Trying to foretell the future is always very dangerous! This is especially true for processor speed; simple predictions related to clock speed do not take into account developments such as pipelining and superscalar architectures. These developments have had an enormous effect on execution speed, and it is this speed that is important, not clock speed.

There is the corollary to Moore's law, named by some as Moore's second law. He points out that as transistor density improves, the cost of the reasearch and production increases dramatically. In the period 1965 to 1995, the costs increased by a factor of about 1000, what was \$12 000 became \$12 million. It may be that the real 'wall', the final limit to CPU execution speed or RAM capacity, will be an economic one, the cost of manufacture may rise to such an extent that components become too expensive. See http://www.imakenews.com/techreview/e_article000003598.cfm.

A relentless increase in the power of the CPU makes this particular component a prime candidate for upgrading a system in order to keep pace with improvements in technology. Figures 1.1 and 1.2 show how the power of the Intel family of processors has increased over the last few decades.

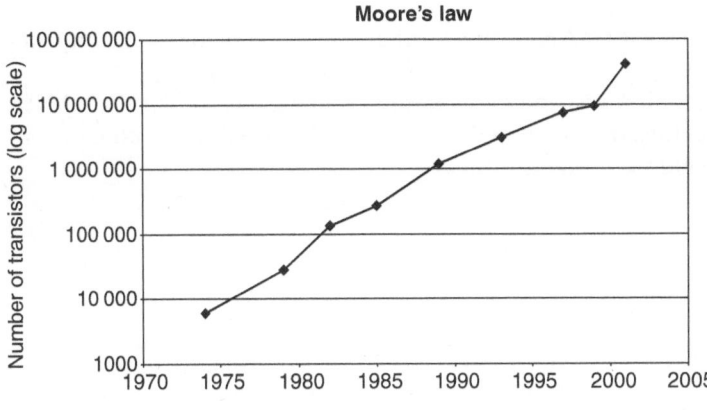

Figure 1.1 *Graph showing number of Pentium microprocessor transistors against date (log scale)*

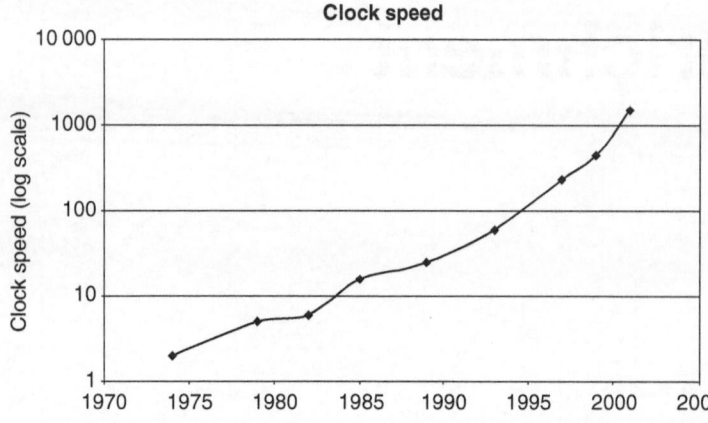

Figure 1.2 *Graph showing clock speeds of Pentium microprocessors against date (log scale)*

Although Moore's law refers to the number of transistors in an integrated circuit, the clock speed of Intel processors seems to conform quite well with the 'law'.

1.2 Enrichment material

Logic and logic gates

The core of a digital computer is logic.

Logic is a branch of mathematics that revolves around the idea of a *proposition*. A proposition is a statement that is either true or false, it has no other possible values.

The proposition 'you are running' may well be not true as you read this page but as a proposition, it has *only* the possibility of being true or false. The argument that you are perhaps sitting down has no value, you are either running or not, the proposition 'you are running' is either true or false. In computers, propositions are translated into 'the circuit is on' or perhaps 'the circuit is at 5 volts', a situation that is easy to build using logic 'gates', circuits that have the property of having only one of two possible states. There is a shorthand for propositions, the use of letters A, B, C etc. For instance proposition A may be 'you are running' and proposition B may be 'it is not raining now'. In this case, B = 0 means it *is* raining because the propositon is false.

It is common to use a '1' for true and '0' for false. (It is becoming common for electrical power switches to be maked 1 for on and 0 for off.)

Logical states can be combined using gates, or dedicated circuits. A circuit that tests if two logical inputs are *both* 'on' or 'true' or '1' (all the same value!) is called an AND gate and will produce an output of 1 only if both the inputs have the value 1, otherwise the circuit will produce a value of 0. This can be summarized on a *truth table* like this:

Input A	Input B	Output
0	0	0
0	1	0
1	0	0
1	1	1

A physical circuit that behaves this way is called an AND *gate*. Such gates can be made in silicon from simpler components such as transistors.

The common logic gates have two inputs and the most common ones have names, i.e. AND, OR, NAND, NOR, XOR etc. and an even simpler one that has one input and one output, the NOT gate (often known as an inverter). The *truth tables* for these are below.

NOT

Input	Output
0	1
1	0

Inputs		Outputs				
A	B	AND	OR	NOR	NAND	XOR
0	0	0	0	1	1	0
0	1	0	1	0	1	1
1	0	0	1	0	1	1
1	1	1	1	0	0	0

NAND means 'Not AND' and you can see from the table that whenever the AND function has the output 1, the NAND function has the output 0 and vice versa. NOR is similar, it has the opposite value to the OR function because it means 'Not OR'. The XOR function means Exclusive OR, the ouput is similar to the OR function except when both inputs = 1.

Circuits can be built using logic gates and to draw these circuits, symbols are used. Unfortunately there are many 'standard' symbols used, but the ones used in Figure 1.3 are at least in common use.

Logic gates are combined to make larger and more complex circuits. A simple example is given in Figure 1.4.

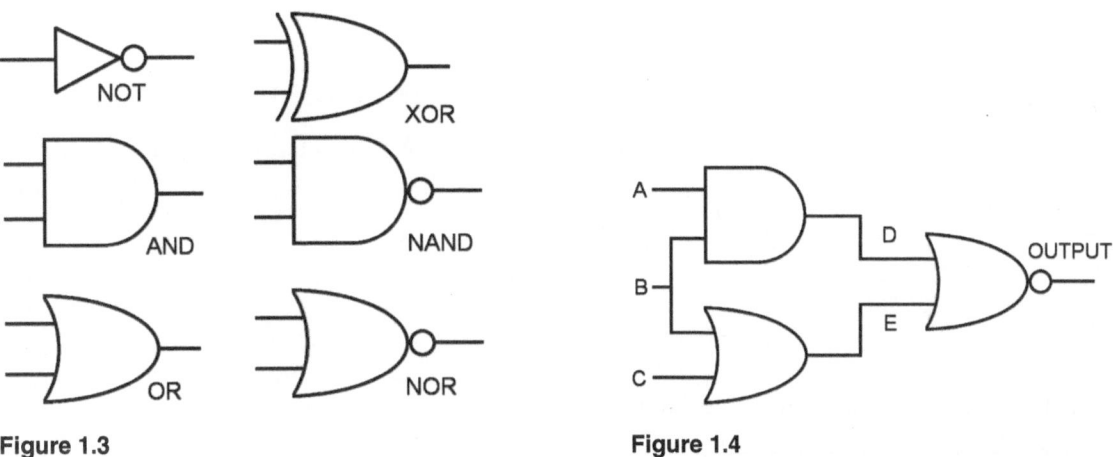

Figure 1.3   **Figure 1.4**

The output from this circuit can be shown on a truth table.

A	B	C	D = A AND B	E = B OR C	OUTPUT = D NOR E
0	0	0	0	0	1
0	0	1	0	1	0
0	1	0	0	1	0
0	1	1	0	1	0
1	0	0	0	0	1
1	0	1	0	1	0
1	1	0	1	1	0
1	1	1	1	1	0

This means the output will only be a 1 if

A = 0 and B = 0 and C = 0

or

A = 1 and B = 0 and C = 0

The circuit has three inputs. Since $2^3 = 8$, there are eight different combinations of 1s and 0s on the three inputs. That is why there are eight lines in the truth table.

Arithmetic with circuits

An addition in decimal is $1 + 1 = 2$ but written in binary is $1 + 1 = 10$. If we set out this in columns we get

```
      1
 +    1
 1    0
```

where the 1 on the left is *carried* over to the next column. This is the notation normally used for addition etc.

If we were to write all the possible additions of two binary digits, we would get

```
      1
 +    1
 1    0
```

```
      1
 +    0
 0    1
```

```
      0
 +    1
 0    1
```

```
      0
 +    0
 0    0
```

These can be summarized on a truth table like this:

Numbers to be added		Outputs	
A	B	Sum	Carry
0	0	0	0
0	1	1	0
1	0	1	0
1	1	0	1

(As we are using binary numbers, they can be treated as logical propositions as they can only have values of 1 or 0.)

You should be able to see that the Sum output has the same truth table as the XOR function and the Carry output has the same truth table as the AND function. This means that we can draw a circuit that will add two binary digits and output the result, i.e. the sum and carry. The resulting circuit is called a *half adder* (see Figure 1.5).

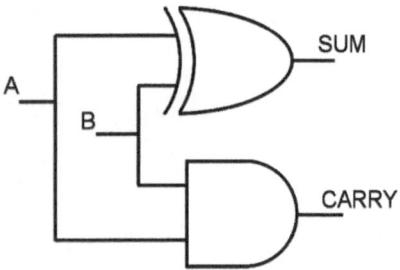

Figure 1.5 *Half adder*

This circuit is fine if all we need to add is just two binary digits but if we need to add say 2 bytes together, we need a circuit that will cope with a carry from the previous column. Consider the addition of 01 and 11:

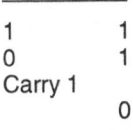

1	1
0	1
Carry 1	
	0

The sum in the right-hand column is $1 + 1 = 0$ but we will need to *carry* 1 over to the next column. If we want a circuit that will add the next column, it will require three inputs, i.e. to add the 1 and 0 and the 1 carried over from the previous column. This can be done by using two half adders joined together as shown in Figure 1.6. The result is called a *full adder*.

We can now use seven full adders and one half adder to add to 8-bit bytes to give an 8-bit result with a carry. In Figure 1.7, each full or half adder is shown as a square symbol:

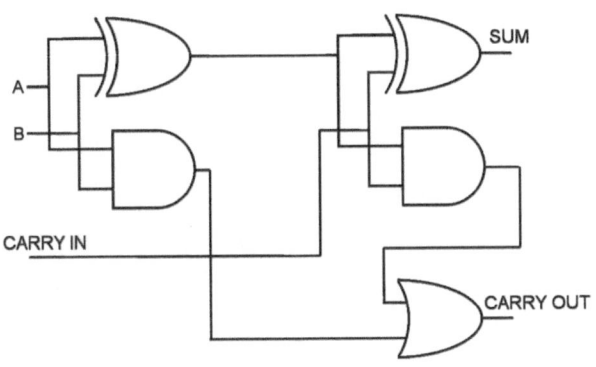

Figure 1.6 *Full adder*

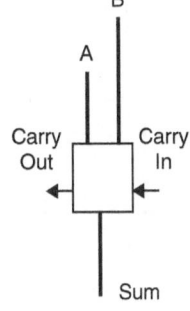

Figure 1.7

These are connected as shown in Figure 1.8.

If we supply the values 147 and 235 as binary values 10010011 and 11101001 we should get the result 382. Eight bits will not hold the value 382 so the result will be 256 (represented as the carry bit) and 126 as the binary pattern 01111110. The carry out from each addition is connected to the carry in of the next adder. This is shown in Figure 1.9.

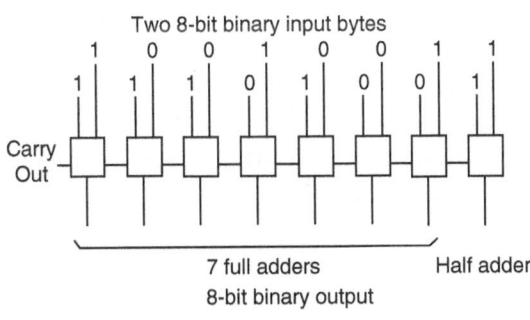

Figure 1.8

Figure 1.9

Address decoder

An 'address' in memory is simply a way of thinking about how memory chips are connected. If a RAM chip has 14 address lines we could place the value 10110100110110 on these lines. If the chip is set to 'read', the data that appears on the data bus is said to be 'at' address 10110100110110. The binary value 10110100110110 is 11 574 in decimal so when 10110100110110 is applied to the chip we can think about the address 11 574.

As $2^{14} = 16\,384$ decimal, the chip can store 16 384 values because that is the number of different addresses that are possible using 14-bit binary. If we need more memory than 16 384 bytes then we could use, say, four chips and connect them with two more 'address' lines. This would give an address bus of 16 lines so a total addressable memory of $2^{16} = 65\,536$ or 64 kb of memory. The circuit in Figure 1.10 shows this arrangement.

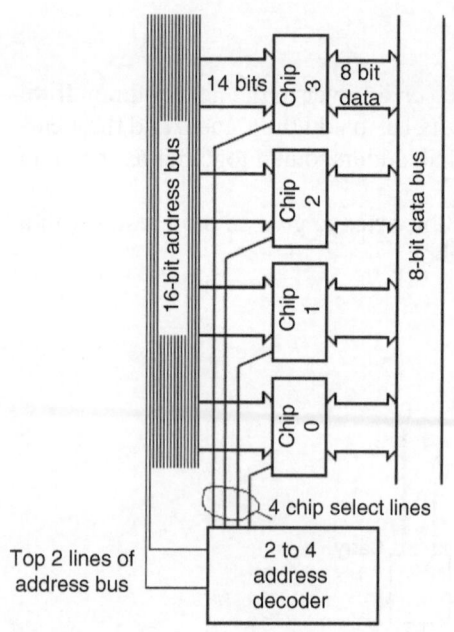

Figure 1.10

The additional two lines are not connected to the chips as they only have 14 lines each, instead they are connected to a two to four address decoder and the four output lines from this circuit are connected to the 'chip enable' lines on the RAM chips. This has the effect of 'turning the chip on' but the decoder is so arranged that only one chip is enabled at any one time.

The truth table for the decoder is simple to work out and is shown below.

Inputs		Outputs			
A	B	Line 3 A AND B	Line 2 A AND NOT(B)	Line 1 NOT(A) AND B	Line 0 NOT(A) AND NOT(B)
0	0	0	0	0	1
0	1	0	0	1	0
1	0	0	1	0	0
1	1	1	0	0	0

The behaviour is that with any value applied to inputs A and B, only one line can be on at once.

The circuit for the decoder is easy to work out. Line 3 simply requires a single AND gate. The other lines are connected to AND and NOT gates as in Figure 1.11.

$\overline{A}$ is the symbol for NOT(A).

Circuits with memory: flip-flops

Circuits such as these show an output almost immediately an input is applied; remember that although circuits are fast, nothing happens instantaneously. When the inputs are changed, the output changes, i.e. they have no memory. If more complex circuits are designed where the outputs *feedback* into the input side, a degree of memory can be achieved. The circuit in Figure 1.12 is called a flip-flop, the output will remain the same until requested to change.

The clock of the RS flip-flop has nothing to do with the gadget on the wall that tells you when it is lunch time! In logic circuits, a clock is simply a series of 1s and 0s otherwise known as a square wave (see Figure 1.13).

When the clock = 0, the output of the left-hand gates both = 1 (see the truth table for a NAND gate). This means the circuit will remain in whatever state it is currently in regardless of the logic on lines S and R. When

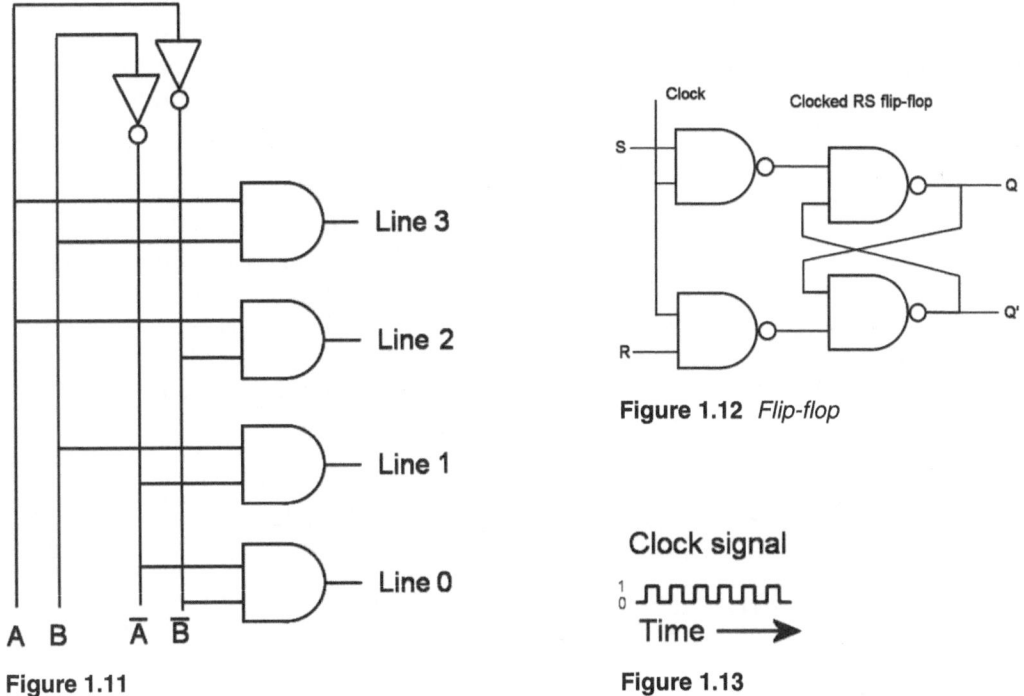

Line 3

Line 2

Line 1

Line 0

A   B $\overline{A}$ $\overline{B}$

Figure 1.11

Clock

Clocked RS flip-flop

S

R

Q

Q'

Figure 1.12 *Flip-flop*

Clock signal

Time ⟶

Figure 1.13

the clock = 1, whatever logic levels are applied to S and R can reach the right-hand gates. If clock = 1, S = 1 and R = 0 the output Q will be Set to 1. If clock = 1, S = 0 and R = 1, the output Q is Reset to 0. You can think of this as an RS or Reset–Set flip-flop. If clock = 1 and S = 0 and R = 0 nothing changes. The other state where clock = 1, S = 1 and R = 1 is not allowed as the output will be indeterminate.

Since S = 1 and R = 1 is not allowed and S = 0 and R = 0 has no effect, a simple modification of the RS flip-flop can be made to eliminate these states. An inverter or NOT gate is inserted between S and R such that when S = 1, R must = 0 and when S = 0, R must = 1. This is known as a D-type flip-flop, its behaviour is very simple, if clock = 1 the circuit 'remembers whatever logic state is applied to D'. If clock = 0, the output remains at what it was, i.e. it is a simple 1-bit memory.

This circuit can be used to store a single binary digit. If eight of these are arranged with a common clock signal, it can be used to store a whole byte of information (see Figure 1.14).

8 inputs

Clock

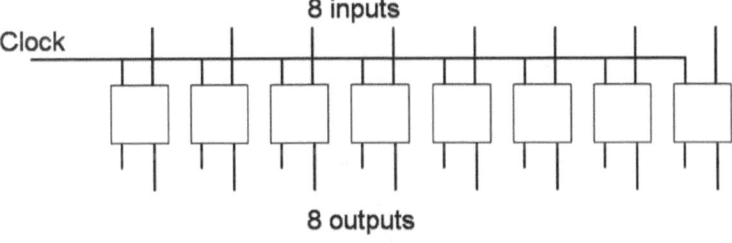

8 outputs

Figure 1.14 *Simple byte store using 8 D-type flip-flops*

1.3 Enrichment material

HEX, binary and other number bases

As with many ideas in mathematics, practice makes perfect. Whilst the *HNC Computing* units do not specifically require fluency with number conversions, most people will find that such fluency is very useful. The idea of the sections presented here is to provide sufficient practice to achieve this fluency.

In 'normal' numbers, the value 137 equals $(1 \times 10^2) + (3 \times 10^1) + (7 \times 10^0)$ or 100 + 30 + 7. Set out as numbers by position we get:

10^2	10^1	10^0
1	3	7

Each column represents the next power of 10. We only use 10 because people are born with 10 fingers, the system works just the same with any positive integer. If we use 8 as the number base (octal numbers), the columns will have the values 8^3, 8^2, 8^1, 8^0 so octal 137 would be $(1 \times 8^2) + (3 \times 8^1) + (7 \times 8^0)$ or, $64 + 24 + 7 = 95$ in decimal. As another example, if we used numbers to the base 12, each column would have values 12^3, 12^2, 12^1, 12^0 so 137 to base 12 in decimal would be $(1 \times 12^2) + (3 \times 12^1) + (7 \times 12^0)$ or $144 + 36 + 7 = 187$. Remember that any number to the power 0 is 1, so $9^0 = 1$ etc.

For a given number base, the same number of symbols is required so in decimal we use the 10 symbols 0–9, in octal we use 0–7 etc. The use of number bases over 10 is sometimes seen as awkward as we have to use letters as number symbols. Hexadecimal (or hex) numbers use 0–9 and A–F, giving a total of 16 symbols. Once you get used to this idea, seeing numbers written as 5DA is not a problem as it simply means $(5 \times 16^2) + (13 \times 16^1) + (10 \times 16^0)$ or $(5 \times 256) + (13 \times 16) + (10 \times 1) = 1498$ in decimal. The trick is to remember that A = 10, B = 11 etc. Binary numbers are simply to the base 2 so we need only 2 symbols, 0 and 1, the system of numbers by position is exactly the same. The binary value $100110 = (1 \times 2^5) + (0 \times 2^4) + (0 \times 2^3) + (1 \times 2^2) + (1 \times 2^1) + (0 \times 2^0)$ or 38 decimal.

Exercise 1

Complete the values in the correct number base. Answer shown at the end of this section.

Binary	Octal	Decimal	Hex
111111011	773		1FB
10100001	241		A1
10111000	270	184	
11000100	304	196	
111001111	717	463	
100111010			13A
110001100			18C
111100001	741		
10011000	230		
	637	415	19F
	6471	3385	D39
			C0C
			209
110110011010			
100010101101			
		1197	
	2132		
	10054		
110011010110			
			98EF
	21657		
		39 780	
			5AAD
11100010110000			
		33 247	

There is no point in using a calculator to do the calculations but using one to check your answers is a good idea. Most calculators will not convert decimal to binary for numbers above 511, this includes Excel 97 using the DEC2BIN function. The calculator that comes with Windows 98, if set to 'Scientific' mode, will work in binary for numbers up to 2^{32} or 4 294 967 295. This is 11111111111111111111111111111111 in binary!

Exercise 2

There are *no* errors in the additions below. What number base is used in each case? Answer shown at the end of this section.

A. 2 + 7 = 10
B. 2 + 4 = 10
C. 325 + 170 = 515
D. 121 + 212 = 1110

Direct conversion of hex to binary to hex

You should practise converting binary to hex and hex to binary directly. Binary to hex is very easy, simply split the binary number into groups of four digits starting at the right-hand side then convert each group into the hex digit.

Example: 101101010000111010 split into groups is 10 1101 0100 0011 1010. This is shown in the table:

10	1101	0100	0011	1010
2	D	4	3	A

so 101101010000111010 binary = 2D43A hex.
  Going the other way is just as easy, write down the binary for each of the hex digits.

3EB5A is 11111010111011010

Hex	3	E	B	5	A
Decimal	3	14	11	5	10
Binary	11	1110	1011	101	1010

Direct conversion octal to binary to octal

The same system as above works for octal numbers, only split the binary into groups of 3 instead of 4.
  (This will work for numbers to base 4, in this case, split the binary into groups of 2. The more general case is to work number bases to integer powers of 2, i.e. 2, 4, 8, 16, 32 etc. To work with number base 32, split into groups of 5, i.e. $2^5 = 32$.)

Conversion using successive division

As shown in the *HNC Computing* book, numbers can be converted by successive division. Here is another example.
  The number 3823 = 111011101111 in binary, looking from the *bottom* of the remainder column.

		Base	Quotient	Remainder
Number =	3823	2	1911	1
Previous quotient	1911	2	955	1
" "	955	2	477	1
" "	477	2	238	1
" "	238	2	119	0
" "	119	2	59	1
" "	59	2	29	1
" "	29	2	14	1
" "	14	2	7	0
" "	7	2	3	1
" "	3	2	1	1
" "	1	2	0	1

The previous table was made using a simple Excel spreadsheet. If the number 3823 is placed in cell A2, the formulae are:

	A	B	C	D
1	Decimal	Base	Quotient	Remainder
2	3823	2	=INT((A2/B2))	=MOD(A2,B2)
3	=C2	2	=INT((A3/B3))	=MOD(A3,B3)
4	=C3	2	=INT((A4/B4))	=MOD(A4,B4)
5	=C4	2	=INT((A5/B5))	=MOD(A5,B5)
6	=C5	2	=INT((A6/B6))	=MOD(A6,B6)
7	=C6	2	=INT((A7/B7))	=MOD(A7,B7)
8	=C7	2	=INT((A8/B8))	=MOD(A8,B8)
9	=C8	2	=INT((A9/B9))	=MOD(A9,B9)
10	=C9	2	=INT((A10/B10))	=MOD(A10,B10)
11	=C10	2	=INT((A11/B11))	=MOD(A11,B11)
12	=C11	2	=INT((A12/B12))	=MOD(A12,B12)
13	=C12	2	=INT((A13/B13))	=MOD(A13,B13)

Simply changing the base value to 8 to use octal numbers, you get from the same spreadsheet:

Decimal	Base	Quotient	Remainder
3823	8	477	7
477	8	59	5
59	8	7	3
7	8	0	7

So 3823 decimal is 7357 octal.

To use this spreadsheet for number bases higher than 10 you would have to wrap the formulae in the right-hand column with the DEC2HEX function like this:

= DEC2HEX(MOD(A2,B2)). Using this you get:

Decimal	Base	Quotient	Remainder
3823	16	238	F
238	16	14	E
14	16	0	E

So 3823 decimal = EEF hex. This gets over any number restrictions in Excel.

Answers
Exercise 1

Binary	Octal	Decimal	Hex
111111011	773	507	1FB
10100001	241	161	A1
10111000	270	184	B8
11000100	304	196	C4
111001111	717	463	1CF
100111010	472	314	13A
110001100	614	396	18C
111100001	741	481	1E1
10011000	230	152	98
110011111	637	415	19F
110100111001	6471	3385	D39
110000001100	6014	3084	C0C
1000001001	1011	521	209
110110011010	6632	3482	D9A
100010101101	4255	2221	8AD
10010101101	2255	1197	4AD
10001011010	2132	1114	45A
1000000101100	10054	4140	102C
110011010110	6326	3286	CD6
1001100011101111	114357	39 151	98EF
10001110101111	21657	9135	23AF
1001101101100100	115544	39 780	9B64
101101010101101	55255	23 213	5AAD
11100010110000	34260	14 512	38B0
1000000111011111	100737	33 247	81DF

Exercise 2
A. 2 + 7 = 10 uses number base 9
B. 2 + 4 = 10 uses number base 6
C. 325 + 170 = 515 uses number base 8 (octal)
D. 121 + 212 = 1110 uses number base 3

1.4 ASCII and Unicode

Whilst the *HNC Computing* units do not specifically require fluency in ASCII, most people will find that such fluency is very useful. The idea of the sections presented here is to provide sufficient practice to achieve this fluency.

Below is the famous part of a poem called *For the Fallen* by Laurence Binyon. A file containing just this text was created using Notepad under Windows.

They shall grow not old, as we that are left grow old;
Age shall not weary them, nor the years condemn.
At the going down of the sun and in the morning
We will remember them.

A hex dump of this file is shown below. The 16 left-hand columns show the hex values of each byte in the file, the right-hand 16 characters show printable text or '.' for non-printable or CTRL characters.

```
09  54  68  65  79  20  73  68  61  6C  6C  20  67  72  6F  77    .They shall grow
20  6E  6F  74  20  6F  6C  64  2C  20  61  73  20  77  65  20    not old, as we
74  68  61  74  20  61  72  65  20  6C  65  66  74  20  67  72    that are left gr
6F  77  20  6F  6C  64  3B  0D  0A  09  41  67  65  20  73  68    ow old;...Age sh
61  6C  6C  20  6E  6F  74  20  77  65  61  72  79  20  74  68    all not weary th
65  6D  2C  20  6E  6F  72  20  74  68  65  20  79  65  61  72    em, nor the year
73  20  63  6F  6E  64  65  6D  6E  2E  0D  0A  09  41  74  20    s condemn....At
74  68  65  20  67  6F  69  6E  67  20  64  6F  77  6E  20  6F    the going down o
66  20  74  68  65  20  73  75  6E  20  61  6E  64  20  69  6E    f the sun and in
20  74  68  65  20  6D  6F  72  6E  69  6E  67  0D  0A  09  57     the morning...W
65  20  77  69  6C  6C  20  72  65  6D  65  6D  62  65  72  20    e will remember
20  74  68  65  6D  2E                                            them.
```

The poem shows a TAB character at the start of each line. TAB is ASCII character 9. Each line ends with a Carriage Return and Line Feed pair. The Carriage Return (CR) is character 13 decimal or 0D hex, Line Feed (LF) is character 10 decimal or 0A hex. After the word 'old' at the end of the first line, you can see the 4 hex bytes 3B 0D 0A 09. These are ';' CR LF and TAB. You can see them in the hex byte section but they cannot be printed in the text section because CR, LF and TAB all cause an action, they do not result in printed characters. That is why they are shown as dots.

Showing ASCII in hex is more efficient than using decimal. In hex, the character A is 41. Any other capital letter can be worked out by using 40 + its position in the alphabet so F is 40 + 6 = 46 hex. Lower case a is 61, i.e. 'A' + 20 so f is 46 + 20 = 66. By following these rules, you can work out any of the letter bytes in the hex dump. A space is character 20.

Exercise

From memory, write down the text and appearance of what would be output on a standard ASCII printer if the following hex bytes were sent to it. Remember the point is only to achieve some fluency. There is little point continuing till the end once you have done this. Answer shown at the end of this section.

57 69 74 68 20 70 72 6F 75 64 20 74 68 61 6E 6B 73 67 69 76 69 6E 67 2C 20 61 20 6D 6F 74 68 65 72 20 66 6F 72 20 68 65 72 20 63 68 69 6C 64 72 65 6E 2C 0D 0A 45 6E 67 6C 61 6E 64 20 6D 6F 75 72 6E 73 20 66 6F 72 20 68 65 72 20 64 65 61 64 20 61 63 72 6F 73 73 20 74 68 65 20 73 65 61 2E

The same poem text as above was saved in a Unicode file using the Unicode option under Wordpad. The resulting hex dump shows what has happened. Each character is stored as 2 bytes because Unicode uses 16 bits per character.

```
FF  FE  09  00  54  00  68  00  65  00  79  00  20  00  73  00    ÿþ..T.h.e.y. .s.
68  00  61  00  6C  00  6C  00  20  00  67  00  72  00  6F  00    h.a.l.l. .g.r.o.
77  00  20  00  6E  00  6F  00  74  00  20  00  6F  00  6C  00    w. .n.o.t. .o.l.
64  00  2C  00  20  00  61  00  73  00  20  00  77  00  65  00    d.,. .a.s. .w.e.
20  00  74  00  68  00  61  00  74  00  20  00  61  00  72  00    .t.h.a.t. .a.r.
65  00  20  00  6C  00  65  00  66  00  74  00  20  00  67  00    e. .l.e.f.t. .g.
72  00  6F  00  77  00  20  00  6F  00  6C  00  64  00  3B  00    r.o.w. .o.l.d.;.
0D  00  0A  00  09  00  41  00  67  00  65  00  20  00  73  00    ......A.g.e. .s.
68  00  61  00  6C  00  6C  00  20  00  6E  00  6F  00  74  00    h.a.l.l. .n.o.t.
20  00  77  00  65  00  61  00  72  00  79  00  20  00  74  00    .w.e.a.r.y. .t.
68  00  65  00  6D  00  2C  00  20  00  6E  00  6F  00  72  00    h.e.m.,. .n.o.r.
20  00  74  00  68  00  65  00  20  00  79  00  65  00  61  00    .t.h.e. .y.e.a.
72  00  73  00  20  00  63  00  6F  00  6E  00  64  00  65  00    r.s. .c.o.n.d.e.
6D  00  6E  00  2E  00  0D  00  0A  00  09  00  41  00  74  00    m.n.........A.t.
20  00  74  00  68  00  65  00  20  00  67  00  6F  00  69  00    .t.h.e. .g.o.i.
6E  00  67  00  20  00  64  00  6F  00  77  00  6E  00  20  00    n.g. .d.o.w.n. .
6F  00  66  00  20  00  74  00  68  00  65  00  20  00  73  00    o.f. .t.h.e. .s.
75  00  6E  00  20  00  61  00  6E  00  64  00  20  00  69  00    u.n. .a.n.d. .i.
6E  00  20  00  74  00  68  00  65  00  20  00  6D  00  6F  00    n. .t.h.e. .m.o.
72  00  6E  00  69  00  6E  00  67  00  0D  00  0A  00  09  00    r.n.i.n.g.......
57  00  65  00  20  00  77  00  69  00  6C  00  6C  00  20  00    W.e. .w.i.l.l. .
72  00  65  00  6D  00  65  00  6D  00  62  00  65  00  72  00    r.e.m.e.m.b.e.r.
20  00  74  00  68  00  65  00  6D  00  2E  00                    .t.h.e.m..
```

Answer to exercise

With proud thanksgiving, a mother for her children,
England mourns for her dead across the sea.

This is the first line of the peom *For the Fallen* by Laurence Binyon used at the start of this section.

1.5 Enrichment material

Concepts of data storage: integer type

A file of Pascal data, 12 bytes long, is known to contain six numbers, the first number has the value 34. What are the other numbers in the file? Attempts to show these numbers directly on the screen using the Windows Notepad program yielded the result:

" ‰ [
íu¡Úÿÿ

which are clearly not numbers if interpreted as ASCII!

In a further attempt to discover the values of these numbers, the DOS DEBUG program was used to provide a hex dump of the file. This yielded the following:

```
0E72:0100  22 00 89 00 5B 0D ED 75-A1 DA FF FF 0D 74 02 EB   "...[..u.....t..
0E72:0110  F8 8B CF 81 E9 82 00 26-88 0E 80 00 34 00 61 0E   .......&....4.a.
0E72:0120  DE BE 10 D4 BA FF FF B8-00 AE CD 2F 3C 00 C3 A0   ........../<...
0E72:0130  D1 E2 0A C0 74 09 56 57-E8 2A 21 5F 5E 73 0A B9   ....t.VW.*!_^s..
0E72:0140  04 01 FC 56 57 F3 A4 5F-5E C3 50 56 33 C9 33 DB   ...VW.._^.PV3.3.
0E72:0150  AC E8 5F 23 74 19 3C 0D-74 15 F6 C7 20 75 06 3A   .._#t.<.t... u.:
0E72:0160  06 02 D3 74 0A 41 3C 22-75 E6 80 F7 20 EB E1 5E   ...t.A<"u... ..^
0E72:0170  58 C3 A1 D7 D7 8B 36 D9-D7 C6 06 1B D9 00 C6 06   X.....6.........
```

(See Appendix B (p. 221) for details of the DEBUG program.)

As the file size was just 12 bytes, it seems obvious that each number is stored in just 2 bytes. The first 12 bytes split into 6 pairs are:

22 00	89 00	5B 0D	ED 75	A1 DA	FF FF

Question

What are the other numbers in the file?

Answer

The first number has the value 34 which is 22 in hex. The byte pair 22 00 seems to agree but is the 'wrong way round'. It would seem a reasonable assumption that the numbers are 16-bit numbers stored low byte:high byte.

Putting the values 'the other way' round yields:

0022
0089
0D5B
75ED
DAA1
FFFF

If these values are now converted to binary we get:

Hex	Binary
0022	0000000000100010
0089	0000000010001001
0D5B	0000110101011011
75ED	0111010111101101
DAA1	1101101010100001
FFFF	1111111111111111

There is now a problem. The left-hand bits of the last two of these numbers is set to 1. There is no way to tell by inspection if the numbers are signed integers, i.e. if this left-hand bit is used as a sign bit or if they are 16-bit unsigned numbers, so the best that is possible is to give both possibilities as shown in the table.

Hex	Binary	Unsigned integer	Signed integer
0022	0000000000100010	34	34
0089	0000000010001001	137	137
0D5B	0000110101011011	3419	3419
75ED	0111010111101101	30189	30189
DAA1	1101101010100001	55969	-9567
FFFF	1111111111111111	65535	-1

Explain the original screen output from Notepad

The result was:

" ‰ [
íu¡Úÿÿ

The file has the 12 bytes:

22 00   89 00   5B 0D   ED 75   A1 DA   FF FF

If these are shown as ASCII characters, all is clear. (See Appendix A on p. 219 for ASCII table.) The file contains hex byte 0D which is a Carriage Return, hence the break in the data after the [character.

Hex	22	00	89	00	5B	0D	ED	75	A1	DA	FF	FF
ASCII	"	NUL	‰	NUL	[	CR	í	u	¡	Ú	ÿ	ÿ

By looking at just a hex dump, there is no way to prove exactly what the data is. The numbers can be decoded as integers provided that it is known the data is made up of integers, the data could well be a set of values in a completely different format. In this case, it is unlikely the data is ASCII as many ASCII files are 'human readable', i.e. you can make out what is in the file by inspection, even if the file contains some control characters.

The only way that data can be assigned real meaning is to open the data in the software used to produce the file or in software that 'knows' what the data is by some other means.

As a further example, you may have recorded the following numerical measurement data in decimal and stored it as unsigned 8-bit bytes:

74	97	109	101	115	32	98	111	110	100

You would then be surprised to find that if you use a piece of software like Word or Notepad, your set of numbers seem to spell a well-known name! You either decode this by hand or use one of the ASCII functions of a spreadsheet.

Little endian numbers

Some systems store 16-bit values as low byte:high byte but others store them as high byte:low byte. When numbers are stored low byte:high byte it is known as 'little endian', i.e. the little number is stored first (or at the lowest address in memory). Systems that store numbers the other way round are known as 'big endian'. This all started in the design of microprocessors so was fixed in the circuitry. Many modern systems are able to deal with both ways of storage.

Worksheet   Unit 1.1 Stored numbers, integers

Six numbers are stored in the first 12 bytes of a file. It is known that they are 16-bit little endian values but it is not known if they are signed or unsigned.

Write down all six numbers as both signed and unsigned decimal values.

```
0E73:0100  B7  00  59  00  68  5D  A3  FE-09  25  19  E3  C3  8B  1E  88   ..Y.h]...%......
0E73:0110  DE  BE  10  D4  BA  FF  FF  B8-00  AE  CD  2F  34  00  62  0E   ............/4.b.
0E73:0120  D1  E2  0A  C0  74  09  56  57-E8  2A  21  5F  5E  73  0A  B9   ....t.VW.*!_^s..
0E73:0130  04  01  FC  56  57  F3  A4  5F-5E  C3  50  56  33  C9  33  DB   ...VW.._^.PV3.3.
0E73:0140  AC  E8  5F  23  74  19  3C  0D-74  15  F6  C7  20  75  06  3A   .._#t.<.t... u.:
0E73:0150  06  02  D3  74  0A  41  3C  22-75  E6  80  F7  20  EB  E1  5E   ...t.A<"u... ..^
0E73:0160  58  C3  A1  D7  D7  8B  36  D9-D7  C6  06  1B  D9  00  C6  06   X.....6.........
0E73:0170  17  D9  00  8B  36  D9  D7  8B-0E  D7  D7  8B  D6  E3  42  51   ....6.........BQ
```

Signed	Unsigned

Worksheet   Unit 1.2 Stored numbers, integers

Six numbers are stored in the first 12 bytes of a file. It is known that they are 16-bit little endian values but it is not known if they are signed or unsigned.

   Write down all six numbers as both signed and unsigned decimal values.

```
0E72:0100   86  6F  BE  72  D4  3B  03  72-44  68  65  10  0D  74  02  EB   .o.r.;.rDhe..t..
0E72:0110   F8  8B  CF  81  E9  82  00  26-88  0E  80  00  34  00  61  0E   .......&....4.a.
0E72:0120   DE  BE  10  D4  BA  FF  FF  B8-00  AE  CD  2F  3C  00  C3  A0   ............/<...
0E72:0130   D1  E2  0A  C0  74  09  56  57-E8  2A  21  5F  5E  73  0A  B9   ....t.VW.*!_^s..
0E72:0140   04  01  FC  56  57  F3  A4  5F-5E  C3  50  56  33  C9  33  DB   ...VW.._^.PV3.3.
0E72:0150   AC  E8  5F  23  74  19  3C  0D-74  15  F6  C7  20  75  06  3A   .._#t.<.t... u.:
0E72:0160   06  02  D3  74  0A  41  3C  22-75  E6  80  F7  20  EB  E1  5E   ...t.A<"u... ..^
0E72:0170   58  C3  A1  D7  D7  8B  36  D9-D7  C6  06  1B  D9  00  C6  06   X.....6.........
```

Signed	Unsigned

Worksheet   Unit 1.3 Stored numbers, integers

Six numbers are stored in the first 12 bytes of a file. It is known that they are 16-bit little endian values but it is not known if they are signed or unsigned.

Write down all six numbers as both signed and unsigned decimal values.

```
0E72:0100   00  16  CE  37  80  29  18  22-DE  32  91  65  0D  74  02  EB   ...7.).'.2.e.t..
0E72:0110   F8  8B  CF  81  E9  82  00  26-88  0E  80  00  34  00  61  0E   .......&....4.a.
0E72:0120   DE  BE  10  D4  BA  FF  FF  B8-00  AE  CD  2F  3C  00  C3  A0   ........../<...
0E72:0130   D1  E2  0A  C0  74  09  56  57-E8  2A  21  5F  5E  73  0A  B9   ....t.VW.*!_^s..
0E72:0140   04  01  FC  56  57  F3  A4  5F-5E  C3  50  56  33  C9  33  DB   ...VW.._^.PV3.3.
0E72:0150   AC  E8  5F  23  74  19  3C  0D-74  15  F6  C7  20  75  06  3A   .._#t.<.t... u.:
0E72:0160   06  02  D3  74  0A  41  3C  22-75  E6  80  F7  20  EB  E1  5E   ...t.A<"u... ..^
0E72:0170   58  C3  A1  D7  D7  8B  36  D9-D7  C6  06  1B  D9  00  C6  06   X.....6.........
```

Signed	Unsigned
	.

Worksheet   Unit 1.4 Stored numbers, integers

Six numbers are stored in the first 12 bytes of a file. It is known that they are 16-bit little endian values but it is not known if they are signed or unsigned.

Write down all six numbers as both signed and unsigned decimal values.

```
0E72:0100   4C  70  4A  2B  BC  65  34  49-6A  66  1D  22  0D  74  02  EB   LpJ+.e4Ijf.".t..
0E72:0110   F8  8B  CF  81  E9  82  00  26-88  0E  80  00  34  00  61  0E   .......&....4.a.
0E72:0120   DE  BE  10  D4  BA  FF  FF  B8-00  AE  CD  2F  3C  00  C3  A0   ............/<...
0E72:0130   D1  E2  0A  C0  74  09  56  57-E8  2A  21  5F  5E  73  0A  B9   ....t.VW.*!_^s..
0E72:0140   04  01  FC  56  57  F3  A4  5F-5E  C3  50  56  33  C9  33  DB   ...VW.._^.PV3.3.
0E72:0150   AC  E8  5F  23  74  19  3C  0D-74  15  F6  C7  20  75  06  3A   .._#t.<.t... u.:
0E72:0160   06  02  D3  74  0A  41  3C  22-75  E6  80  F7  20  EB  E1  5E   ...t.A<"u... ..^
0E72:0170   58  C3  A1  D7  D7  8B  36  D9-D7  C6  06  1B  D9  00  C6  06   X.....6.........
```

Signed	Unsigned

Worksheet   Unit 1.5 Stored numbers, integers

Six numbers are stored in the first 12 bytes of a file. It is known that they are 16-bit little endian values but it is not known if they are signed or unsigned.

Write down all six numbers as both signed and unsigned decimal values.

```
0E72:0100   17 59 AE 64 36 28 6D 1E-30 0E 21 70 0D 74 02 EB   .Y.d6(m.0.!p.t..
0E72:0110   F8 8B CF 81 E9 82 00 26-88 0E 80 00 34 00 61 0E   .......&....4.a.
0E72:0120   DE BE 10 D4 BA FF FF B8-00 AE CD 2F 3C 00 C3 A0   ........../<...
0E72:0130   D1 E2 0A C0 74 09 56 57-E8 2A 21 5F 5E 73 0A B9   ....t.VW.*!_^s..
0E72:0140   04 01 FC 56 57 F3 A4 5F-5E C3 50 56 33 C9 33 DB   ...VW.._^.PV3.3.
0E72:0150   AC E8 5F 23 74 19 3C 0D-74 15 F6 C7 20 75 06 3A   .._#t.<.t... u.:
0E72:0160   06 02 D3 74 0A 41 3C 22-75 E6 80 F7 20 EB E1 5E   ...t.A<"u... ..^
0E72:0170   58 C3 A1 D7 D7 8B 36 D9-D7 C6 06 1B D9 00 C6 06   X.....6.........
```

Signed	Unsigned

Worksheet   Unit 1.6 Stored numbers, integers

Six numbers are stored in the first 12 bytes of a file. It is known that they are 16-bit little endian values but it is not known if they are signed or unsigned.

Write down all six numbers as both signed and unsigned decimal values.

```
0E72:0100   C7  8D  C7  9C  73  3B  30  3C-36  FD  8A  CF  0D  74  02  EB    ....s;0<6....t..
0E72:0110   F8  8B  CF  81  E9  82  00  26-88  0E  80  00  34  00  61  0E    .......&....4.a.
0E72:0120   DE  BE  10  D4  BA  FF  FF  B8-00  AE  CD  2F  3C  00  C3  A0    .........../<...
0E72:0130   D1  E2  0A  C0  74  09  56  57-E8  2A  21  5F  5E  73  0A  B9    ....t.VW.*!_^s..
0E72:0140   04  01  FC  56  57  F3  A4  5F-5E  C3  50  56  33  C9  33  DB    ...VW.._^.PV3.3.
0E72:0150   AC  E8  5F  23  74  19  3C  0D-74  15  F6  C7  20  75  06  3A    .._#t.<.t... u.:
0E72:0160   06  02  D3  74  0A  41  3C  22-75  E6  80  F7  20  EB  E1  5E    ...t.A<"u... ..^
0E72:0170   58  C3  A1  D7  D7  8B  36  D9-D7  C6  06  1B  D9  00  C6  06    X.....6.........
```

Signed	Unsigned

Worksheet   Unit 1.7 Stored numbers, integers

Six numbers are stored in the first 12 bytes of a file. It is known that they are 16-bit little endian values but it is not known if they are signed or unsigned.

Write down all six numbers as both signed and unsigned decimal values.

```
0E72:0100   EA  CB  25  1A  8C  07  9D  C6-DD  4C  1F  BD  0D  74  02  EB   ..%......L...t..
0E72:0110   F8  8B  CF  81  E9  82  00  26-88  0E  80  00  34  00  61  0E   .......&....4.a.
0E72:0120   DE  BE  10  D4  BA  FF  FF  B8-00  AE  CD  2F  3C  00  C3  A0   .........../<...
0E72:0130   D1  E2  0A  C0  74  09  56  57-E8  2A  21  5F  5E  73  0A  B9   ....t.VW.*!_^s..
0E72:0140   04  01  FC  56  57  F3  A4  5F-5E  C3  50  56  33  C9  33  DB   ...VW.._^.PV3.3.
0E72:0150   AC  E8  5F  23  74  19  3C  0D-74  15  F6  C7  20  75  06  3A   .._#t.<.t... u.:
0E72:0160   06  02  D3  74  0A  41  3C  22-75  E6  80  F7  20  EB  E1  5E   ...t.A<"u... ..^
0E72:0170   58  C3  A1  D7  D7  8B  36  D9-D7  C6  06  1B  D9  00  C6  06   X.....6.........
```

Signed	Unsigned

Worksheet   Unit 1.8 Stored numbers, integers

Six numbers are stored in the first 12 bytes of a file. It is known that they are 16-bit little endian values but it is not known if they are signed or unsigned.

Write down all six numbers as both signed and unsigned decimal values.

```
0E72:0100  8C 71 58 4A 8C 52 A3 8B-F8 5B 0D C6 0D 74 02 EB  .qXJ.R...[...t..
0E72:0110  F8 8B CF 81 E9 82 00 26-88 0E 80 00 34 00 61 0E  .......&....4.a.
0E72:0120  DE BE 10 D4 BA FF FF B8-00 AE CD 2F 3C 00 C3 A0  ............/<...
0E72:0130  D1 E2 0A C0 74 09 56 57-E8 2A 21 5F 5E 73 0A B9  ....t.VW.*!_^s..
0E72:0140  04 01 FC 56 57 F3 A4 5F-5E C3 50 56 33 C9 33 DB  ...VW.._^.PV3.3.
0E72:0150  AC E8 5F 23 74 19 3C 0D-74 15 F6 C7 20 75 06 3A  .._#t.<.t... u.:
0E72:0160  06 02 D3 74 0A 41 3C 22-75 E6 80 F7 20 EB E1 5E  ...t.A<"u... ..^
0E72:0170  58 C3 A1 D7 D7 8B 36 D9-D7 C6 06 1B D9 00 C6 06  X.....6.........
```

Signed	Unsigned

Worksheet   Unit 1.9 Stored numbers, integers

Six numbers are stored in the first 12 bytes of a file. It is known that they are 16-bit little endian values but it is not known if they are signed or unsigned.

Write down all six numbers as both signed and unsigned decimal values.

```
0E72:0100   6B  28  FF  64  2B  13  1C  E4-A9  BA  2A  00  0D  74  02  EB   k(.d+.....*..t..
0E72:0110   F8  8B  CF  81  E9  82  00  26-88  0E  80  00  34  00  61  0E   .......&....4.a.
0E72:0120   DE  BE  10  D4  BA  FF  FF  B8-00  AE  CD  2F  3C  00  C3  A0   ........../<...
0E72:0130   D1  E2  0A  C0  74  09  56  57-E8  2A  21  5F  5E  73  0A  B9   ....t.VW.*!_^s..
0E72:0140   04  01  FC  56  57  F3  A4  5F-5E  C3  50  56  33  C9  33  DB   ...VW.._^.PV3.3.
0E72:0150   AC  E8  5F  23  74  19  3C  0D-74  15  F6  C7  20  75  06  3A   .._#t.<.t... u.:
0E72:0160   06  02  D3  74  0A  41  3C  22-75  E6  80  F7  20  EB  E1  5E   ...t.A<"u... ..^
0E72:0170   58  C3  A1  D7  D7  8B  36  D9-D7  C6  06  1B  D9  00  C6  06   X.....6.........
```

Signed	Unsigned

Worksheet   Unit 1.10 Stored numbers, integers

Six numbers are stored in the first 12 bytes of a file. It is known that they are 16-bit little endian values but it is not known if they are signed or unsigned.

Write down all six numbers as both signed and unsigned decimal values.

```
0E72:0100   80  8C  03  5F  FB  16  1F  07-12  6E  AC  02  0D  74  02  EB    ..._.....n...t..
0E72:0110   F8  8B  CF  81  E9  82  00  26-88  0E  80  00  34  00  61  0E    .......&....4.a.
0E72:0120   DE  BE  10  D4  BA  FF  FF  B8-00  AE  CD  2F  3C  00  C3  A0    .........../<...
0E72:0130   D1  E2  0A  C0  74  09  56  57-E8  2A  21  5F  5E  73  0A  B9    ....t.VW.*!_^s..
0E72:0140   04  01  FC  56  57  F3  A4  5F-5E  C3  50  56  33  C9  33  DB    ...VW.._^.PV3.3.
0E72:0150   AC  E8  5F  23  74  19  3C  0D-74  15  F6  C7  20  75  06  3A    .._#t.<.t... u.:
0E72:0160   06  02  D3  74  0A  41  3C  22-75  E6  80  F7  20  EB  E1  5E    ...t.A<"u... ..^
0E72:0170   58  C3  A1  D7  D7  8B  36  D9-D7  C6  06  1B  D9  00  C6  06    X.....6.........
```

Signed	Unsigned

Concepts of data storage: floating point type

Floating point numbers are very useful for the convenient representation of fractions but they are not always accurate.

Generally, decimal numbers whose fixed point binary presentation fits inside the number of bytes used by the floating point format will be accurate, i.e. there will be no loss of bits. Numbers such as 0.01 do not fit into this category as 0.01 decimal in fixed point binary is approximately

0.0000001010001111010111000010100011110101110000101000000011110110 to 64 binary places or
0.000000101000111101011100001010001111010111000010 to 48 binary places.

If the floating point representation used, say, 48 bits, the last $64 - 48 = 16$ bits would be lost, i.e. the number is less accurate. Although this is true, it is important to get it in perspective. The difference, in decimal, between a 48-bit and 64-bit representation of decimal 0.01 is just 1.810^{-15}!

Applications where large numbers of iterations are performed are likely to show errors. An iteration is an operation that is done over and over again.

Consider this loop in Pascal, assuming the variable x is of type real:

```
x := 0;
repeat
     something(x);
     {call procedure called 'something'.
     What it is, is of no importance here}
     x := x+0.001;
until x = 1000;
```

This should execute 1 000 000 times but the loop may well not actually finish, it may go on forever (i.e. it may be an infinite loop). This is because x may never equal 1000, it will simply be very close. The 'bug' is fixed by using the condition until $x \geqslant 1000$, this will terminate.

The better way to write this loop would be to use :=

```
for i := 0 to 1000000 do
     begin
          x := i/1000;
          something(x);
     end;
```

where i is of type *longint*. The loop will execute a known number of times and there is no cumulative error. Although the floating point calculation `x := i/1000;` will produce an error, this error will always be very small. The previous example added any error 1 000 000 times so in effect 'amplifying' the error.

Longints can be from $-2\,147\,483\,648$ to $2\,147\,483\,647$ because they are signed 32-bit numbers.

Worksheet   Unit 1.11 Floating point fractional numbers

You can convert a decimal fraction to a binary fraction to any number of places. The method is to successively multiply by 2 and to take the integer or whole part of the answer as a binary digit.

For example, convert 0.5703125 to a fixed point binary fraction.

0.5703125	× 2 =	1.140625	integer part =	1
0.140625	× 2 =	0.28125	integer part =	0
0.28125	× 2 =	0.5625	integer part =	0
0.5625	× 2 =	1.125	integer part =	1
0.125	× 2 =	0.25	integer part =	0
0.25	× 2 =	0.5	integer part =	0
0.5	× 2 =	1	integer part =	1

Answer = 0.1001001 reading from the top

Use a spreadsheet to do the conversions of the following numbers:

* to at least 60 binary places;
* if the numbers were then to be used in a floating point number with a 48-bit mantissa, show which of the numbers would cause a slight inaccuracy.

Decimal	60 place binary	48-bit accurate?
0.1		
0.25678		
0.0625		
0.251220703		
0.141592654		

Worksheet   Unit 1.12 Number bases

Fill in the spaces in the table by converting number bases as required.

Hints:
- Binary to hex or hex to binary, use groups of four binary digits.
- Binary to octal or octal to binary, use groups of three binary digits.
- If running Windows, you can use Start > Programs > Accessories > Calculator (with View set to Scientific) to check your answers. There is no point whatever using the calculator to complete the conversions, you should do them on paper alone to get the most practice.

Decimal	Hex	Octal	Binary
	1C9DC		
			010110010111
		2024	
4			
91 509			
		2340	
	45		
	18A0B		
	154		
66 182			
			0010001000001000
10 815			

Worksheet   Unit 1.13 Number bases

Fill in the spaces in the table by converting number bases as required.

Hints:
- Binary to hex or hex to binary, use groups of four binary digits.
- Binary to octal or octal to binary, use groups of three binary digits.
- If running Windows, you can use Start > Programs > Accessories > Calculator (with View set to Scientific) to check your answers. There is no point whatever using the calculator to complete the conversions, you should do them on paper alone to get the most practice.

Decimal	Hex	Octal	Binary
			0001000100100000
4252			
		264767	
47			
			0010011111110101
		350	
589			
			0110100010110101
	DD		
			110001011000
	226D		
		675	

Worksheet   Unit 1.14 Number bases

Fill in the spaces in the table by converting number bases as required.

Hints:
- Binary to hex or hex to binary, use groups of four binary digits.
- Binary to octal or octal to binary, use groups of three binary digits.
- If running Windows, you can use Start > Programs > Accessories > Calculator (with View set to Scientific) to check your answers. There is no point whatever using the calculator to complete the conversions, you should do them on paper alone to get the most practice.

Decimal	Hex	Octal	Binary
		24040	
			0001000111110111
		24104	
		20066	
377			
			101110000011
			0001100100101011
11 389			
9148			
	285C		
	1F0A3		
	D97		

Worksheet   Unit 1.15 Number bases

Fill in the spaces in the table by converting number bases as required.

Hints:

- Binary to hex or hex to binary, use groups of four binary digits.
- Binary to octal or octal to binary, use groups of three binary digits.
- If running Windows, you can use Start > Programs > Accessories > Calculator (with View set to Scientific) to check your answers. There is no point whatever using the calculator to complete the conversions, you should do them on paper alone to get the most practice.

Decimal	Hex	Octal	Binary
	478E		
	1F0C		
	2086		
			0001100100101011
		2001	
93 208			
98 757			
		11176	
			1101101000100101
			0010110010111101
			0001110101010001
			11010010

Worksheet   Unit 1.16 Number bases

Fill in the spaces in the table by converting number bases as required.

Hints:
- Binary to hex or hex to binary, use groups of four binary digits.
- Binary to octal or octal to binary, use groups of three binary digits.
- If running Windows, you can use Start > Programs > Accessories > Calculator (with View set to Scientific) to check your answers. There is no point whatever using the calculator to complete the conversions, you should do them on paper alone to get the most practice.

Decimal	Hex	Octal	Binary
	15E		
	297		
	1FE		
			011010101010
			0010001000100110
			0001000111001110
8702			
8458			
12 469			
		10260	
		115	
		12542	

Worksheet   Unit 1.17 Number bases

Fill in the spaces in the table by converting number bases as required.

Hints:
- Binary to hex or hex to binary, use groups of four binary digits.
- Binary to octal or octal to binary, use groups of three binary digits.
- If running Windows, you can use Start > Programs > Accessories > Calculator (with View set to Scientific) to check your answers. There is no point whatever using the calculator to complete the conversions, you should do them on paper alone to get the most practice.

Decimal	Hex	Octal	Binary
	3087		
	21EB		
	123B5		
112 061			
			000101100111
			11101001
2973			
123 366			
			00011101
		25032	
		1422	
		23667	

Simple files: example of bitmapped graphics files

The BTEC unit specifies that students should understand simple files and the basic concepts and principles involved in the storage of data. A single bitmapped file is presented here to show the contents and file structure.

This picture is stored in a common file format called a 'Bitmap'. (Under the Windows operating system, such files will have the file extension .BMP.)

Figure 1.15

The file is 54 by 76 pixels and was drawn with 256 colours. This means that each pixel is stored in 1 byte in the file so there are $54 \times 76 = 4104$ pixels in the whole image. The file is 5334 bytes in size, the rest of the $5334 - 4104$ bytes of data are stored in a 'file header' and contains mainly a 'colour palette', a means to specify which colours are to be used.

A typical system will use 3 bytes to store each pixel, one for the red intensity, one for green and one for blue. Any of 16.7 million colours can then be made as 3 bytes = 24 bits and $2^{24} = 16.7$ million.

In an image like this one that can store only 256 colours per pixel, each pixel is represented by 1 byte and the 'palette' holds 256 entries, each of the 256 entries holds a 3-byte colour. For example, to find the actual colour of 'colour number' 38 in this file, you would look at entry number 38 in the palette; there you would find 3 bytes, one for red, one for green and one for blue, the actual colour of the pixel. Although the colour of any pixel can be one of 16.7 million colours, in a 256 colour file like this one, you can only have 256 choices from the 16.7 million colours.

The maximum number of colours in a bitmapped file are shown in the table below.

Bits per pixel	Maximum number of colours	Known as
2	$2^2 = 4$	
4	$2^4 = 16$	
8	$2^8 = 256$	
16	$2^{16} = 65\,536$	Hi-Color
24	$2^{24} = 16\,777\,216$	Tru-Color or 24-bit color

The file *header* is in two parts, the *bitmap file header* and the *bitmap information header*. These give information such as image width and height, number of colours per pixel etc. as shown in the table below. The 'start position' refers to the byte position in the file so, for example, if you took 19 bytes into the file you will find the width of the image in pixels, shown in hex, stored *little endian*, i.e. with the least significant byte (or little) byte stored first.

This image is 54 pixels wide which is 36 hex. Starting from byte position 19, the file contains the 4 bytes 36 00 00 00. Because this is little endian, turn it round to *big endian format* to give 00 00 00 36, the image width. If the byte sequence were 23 A4 3D 00, the file size would be 3DA423 or 4 039 715 bytes.

In a similar way, looking at byte position 29, you get the 2 bytes 08 00; turned round, this gives the value 8 bits or 1 byte per pixel, correct for a 256 colour image. At byte position 11 you get 36 04 00 00; turned round to big endian format this gives 436 (hex) or 1078. This is the offset to the bitmap data, i.e. the image data starts 1078 bytes into the file. Oddly, the image data is stored back to front, the last row of data is the first row of pixels in the image.

Bitmap image file format details

Start position	Size in bytes	Meaning or use
The bitmap file header		
1	2	set to 'BM' to show this is a .bmp-file.
3	4	size of the file in bytes.
7	2	always be set to zero.
9	2	always be set to zero.
11	4	offset from the beginning of the file to the bitmap data.

(continued)

Start position	Size in bytes	Meaning or use
The bitmap information header		
15	4	size of the bitmap information header structure, in bytes.
19	4	width of the image, in pixels.
23	4	height of the image, in pixels.
27	2	set to zero.
29	2	number of bits per pixel.
31	4	compression, usually set to zero, 0 = no compression.
35	4	size of the image data, in bytes.
39	4	horizontal pixels per metre, usually set to zero.
43	4	vertical pixels per metre, usually set to zero.
47	4	number of colours used in the bitmap.
51	4	number of colours that are 'important' for the bitmap, if set to zero, all colours are important.

The bit map data

The data that follows has a format that depends on the information in the header. If the file is a 24-bit file (16.7 million colours), there are 3 bytes per pixel, one for red intensity, one for green and one for blue. There is no palette.

If the file is 1 byte per pixel (256 colours), each pixel is just one value that is used with the palette to determine the colour of the pixel.

Figure 1.16

Hex dump of image of man1.bmp

File position of first item in row	File contents (hex)
1	42 4D D6 14 00 00 00 00-00 00 36 04 00 00 28 00
17	00 00 36 00 00 00 4C 00-00 00 01 00 08 00 00 00
33	00 00 A0 10 00 00 00 0B-00 00 00 0B 00 00 00 01
49	00 00 00 01 00 00 10 08-10 00 18 10 18 00 21 18
65	18 00 21 21 18 00 29 21-18 00 29 29 18 00 21 18
81	21 00 29 21 21 00 31 21-21 00 29 29 21 00 31 29
97	21 00 39 29 21 00 31 31-21 00 39 31 21 00 39 39
113	21 00 31 29 21 00 39 29-29 00 31 31 29 00 39 31
129	29 00 42 31 29 00 39 39-29 00 42 39 29 00 4A 42
145	29 00 31 31 31 00 39 31-31 00 42 31 31 00 39 39
161	31 00 42 39 31 00 4A 39-31 00 42 42 31 00 4A 42
177	31 00 4A 4A 31 00 52 4A-31 00 31 31 39 00 42 39
193	39 00 4A 39 39 00 42 42-39 00 4A 42 39 00 52 42
209	39 00 52 4A 39 00 5A 52-39 00 4A 42 42 00 52 42
225	42 00 4A 4A 42 00 52 4A-42 00 52 52 42 00 5A 52
241	42 00 63 5A 42 00 42 42-4A 00 4A 42 4A 00 52 42
257	4A 00 4A 4A 4A 00 52 4A-4A 00 5A 4A 4A 00 52 52
273	4A 00 5A 52 4A 00 5A 5A-4A 00 63 5A 4A 00 5A 52
289	52 00 63 52 52 00 5A 5A-52 00 63 5A 52 00 63 63
305	52 00 52 52 5A 00 5A 52-5A 00 5A 5A 5A 00 63 5A
321	5A 00 6B 5A 5A 00 63 63-5A 00 6B 63 5A 00 6B 6B
337	5A 00 73 6B 5A 00 5A 52-63 00 63 5A 63 00 63 63
353	63 00 6B 63 63 00 6B 6B-63 00 73 6B 63 00 73 73
369	63 00 63 63 6B 00 6B 63-6B 00 6B 6B 6B 00 73 6B
385	6B 00 73 73 6B 00 7B 73-6B 00 84 7B 6B 00 6B 63
	Some data deleted here to save space
1009	FF 00 FF FF FF 00 FF FF-FF 00 FF FF FF 00 FF FF
1025	FF 00 FF FF FF 00 FF FF-FF 00 FF FF FF 00 FF FF
1041	FF 00 FF FF FF 00 FF FF-FF 00 FF FF FF 00 FF FF

(continued)

File position of first item in row	File contents (hex)
1057	FF 00 FF FF FF 00 FF FF-FF 00 FF FF FF 00 FF FF
1073	FF 00 FF FF FF 00 BB BB-BB BB BB BB BB BB BB BB
1089	BB BB BB BB BB BB BB BB-BB BB BB BB BB BB BB BB
1105	BB BB BB BB BB BB BB BB-BB BB BB BB BB BB BB BB
1121	BB BB BB BB BB BB BB BB-BB BB BB BB 00 00 BB BB
1137	BB BB BB BB BB BB BB BB-BB BB BB BB BB BB BB BB
1153	BB BB BB BB BB BB BB BB-BB BB BB BB BB BB BB BB
1169	BB BB BB BB BB BB BB BB-BB BB BB BB BB BB BB BB
1185	BB BB BB BB 00 00 BB BB-BB BB BB BB BB BB BB BB
1201	BB BB BB BB BB BB BB BB-BB BB BB BB BB BB BB BB
1217	BB BB BB BB BB BB BB BB-BB BB BB BB BB BB BB BB
1233	BB BB BB BB BB BB BB BB-BB BB BB BB 00 00 BB BB
1249	BB BB BB BB BB BB BB BB-BB BB BB BB BB BB BB BB
1265	BB BB BB BB BB BB BB BB-BB BB BB BB BB BB BB BB
1281	BB BB BB BB BB BB BB BB-BB BB BB BB BB BB BB BB
1297	BB BB BB BB 00 00 BB BB-BB BB BB BB BB BB BB BB
	Some data deleted here to save space
5201	BB BB BB BB BB BB BB BB-BB BB BB BB BB BB BB BB
5217	BB BB BB BB 00 00 BB BB-BB BB BB BB BB BB BB BB
5233	BB BB BB BB BB BB BB BB-BB BB BB BB BB BB BB BB
5249	BB BB BB BB BB BB BB BB-BB BB BB BB BB BB BB BB
5265	BB BB BB BB BB BB BB BB-BB BB BB BB 00 00 BB BB
5281	BB BB BB BB BB BB BB BB-BB BB BB BB BB BB BB BB
5297	BB BB BB BB BB BB BB BB-BB BB BB BB BB BB BB BB
5313	BB BB BB BB BB BB BB BB-BB BB BB BB BB BB BB BB
5329	BB BB BB BB 00 00

Worksheet   Unit 1.18 Storage of data, bitmaps

The hex dump below is the top of a bitmapped file. Look at the values in the file header and fill in the correct values in the table.

Bitmap image file format details

Start position	Size in bytes	Meaning or use
The bitmap file header		
1	2	set to 'BM' to show this is a .bmp-file.
3	4	size of the file in bytes.
7	2	always be set to zero.
9	2	always be set to zero.
11	4	offset from the beginning of the file to the bitmap data.
The bitmap information header		
15	4	size of the bitmap information header structure, in bytes.
19	4	width of the image, in pixels.
23	4	height of the image, in pixels.
27	2	set to zero.
29	2	number of bits per pixel.
31	4	compression, usually set to zero, 0 = no compression.
35	4	size of the image data, in bytes.
39	4	horizontal pixels per metre, usually set to zero.
43	4	vertical pixels per metre, usually set to zero.
47	4	number of colours used in the bitmap.
51	4	number of colours that are 'important' for the bitmap, if set to zero, all colours are important.

```
42  4D  36  1B  B7  00  00  00  00  00  36  00  00  00  28  00    BM6.·......6...(.
00  00  D0  07  00  00  D0  07  00  00  01  00  18  00  00  00    ..Đ...Đ.........
00  00  00  1B  B7  00  66  5C  00  00  66  5C  00  00  00  00    .....·.f\..f\....
00  00  00  00  00  00  FF  FF  FF  FF  FF  FF  FF  FF  FF  FF    ......ÿÿÿÿÿÿÿÿÿÿ
FF  FF  FF  FF  FF  FF  FF  FF  FF  FF  FF  FF  FF  FF  FF  FF    ÿÿÿÿÿÿÿÿÿÿÿÿÿÿÿÿ
FF  FF  FF  FF  FF  FF  FF  FF  FF  FF  FF  FF  FF  FF  FF  FF    ÿÿÿÿÿÿÿÿÿÿÿÿÿÿÿÿ
FF  FF  FF  FF  FF  FF  FF  FF  FF  FF  FF  FF  FF  FF  FF  FF    ÿÿÿÿÿÿÿÿÿÿÿÿÿÿÿÿ
FF  FF  FF  FF  FF  FF  FF  FF  FF  FF  FF  FF  FF  FF  FF  FF    ÿÿÿÿÿÿÿÿÿÿÿÿÿÿÿÿ
FF  FF  FF  FF  FF  FF  FF  FF  FF  FF  FF  FF  FF  FF  FF  FF    ÿÿÿÿÿÿÿÿÿÿÿÿÿÿÿÿ
FF  FF  FF  FF  FF  FF  FF  FF  FF  FF  FF  FF  FF  FF  FF  FF    ÿÿÿÿÿÿÿÿÿÿÿÿÿÿÿÿ
FF  FF  FF  FF  FF  FF  FF  FF  FF  FF  FF  FF  FF  FF  FF  FF    ÿÿÿÿÿÿÿÿÿÿÿÿÿÿÿÿ
FF  FF  FF  FF  FF  FF  FF  FF  FF  FF  FF  FF  FF  FF  FF  FF    ÿÿÿÿÿÿÿÿÿÿÿÿÿÿÿÿ
```

In decimal, what is the:

	Hex values, little endian	Hex values, big endian	Decimal value
Image file size			
Image width			
Image height			
Number of bits per pixel			
Max. number of colours possible	n/a	n/a	

Worksheet   Unit 1.19 Storage of data, bitmaps

The hex dump below is the top of a bitmapped file. Look at the values in the file header and fill in the correct values in the table.

Bitmap image file format details

Start position	Size in bytes	Meaning or use
The bitmap file header		
1	2	set to 'BM' to show this is a .bmp-file.
3	4	size of the file in bytes.
7	2	always be set to zero.
9	2	always be set to zero.
11	4	offset from the beginning of the file to the bitmap data.
The bitmap information header		
15	4	size of the bitmap information header structure, in bytes.
19	4	width of the image, in pixels.
23	4	height of the image, in pixels.
27	2	set to zero.
29	2	number of bits per pixel.
31	4	compression, usually set to zero, 0 = no compression.
35	4	size of the image data, in bytes.
39	4	horizontal pixels per metre, usually set to zero.
43	4	vertical pixels per metre, usually set to zero.
47	4	number of colours used in the bitmap.
51	4	number of colours that are 'important' for the bitmap, if set to zero, all colours are important.

```
42  4D  8E  37  38  00  00  00  00  00  36  04  00  00  28  00   BM□78.....6...(.
00  00  22  09  00  00  26  06  00  00  01  00  08  00  00  00   .."...&.........
00  00  58  33  38  00  F2  1E  00  00  F2  1E  00  00  00  01   ..X38.ð...ð.....
00  00  00  01  00  00  45  47  59  00  7C  4D  55  00  50  76   ......EGY.|MU.Pv
66  00  84  75  5B  00  37  65  86  00  6C  64  84  00  82  59   f.„u[.7et.ld„.,Y
87  00  84  7C  81  00  27  67  A5  00  29  8A  A3  00  53  5B   ‡.„|□.'g¥.)Š£.S[
A7  00  53  87  A4  00  73  54  A7  00  74  6C  AA  00  77  80   §.S‡¤.sT§.tlª.w□
94  00  75  81  B0  00  A8  59  45  00  B3  79  4B  00  A7  5D   ".u□°."YE.³yK.§]
6E  00  B0  7A  6B  00  9A  67  7D  00  B8  6B  7E  00  A3  83   n.°zk.šg}.,k~.£ƒ
78  00  C2  7F  78  00  8E  68  8E  00  8C  6C  AB  00  AB  6E   x.Â□x.□h□.Œl«.«n
8D  00  A5  73  A6  00  98  88  8D  00  95  88  AB  00  BD  84   □.¥s|.˜^□.•^«.¹⁄₂„
84  00  B9  84  A4  00  27  97  A1  00  32  B5  A4  00  5A  A7   „.¹„¤.'—¡.2µ¤.Z§
```

In decimal, what is the:

	Hex values, little endian	Hex values, big endian	Decimal value
Image file size			
Image width			
Image height			
Number of bits per pixel			
Max. number of colours possible	n/a	n/a	

Worksheet   Unit 1.20 Storage of data, bitmaps

The hex dump below is the top of a bitmapped file. Look at the values in the file header and fill in the correct values in the table.

Bitmap image file format details

Start position	Size in bytes	Meaning or use
The bitmap file header		
1	2	set to 'BM' to show this is a .bmp-file.
3	4	size of the file in bytes.
7	2	always be set to zero.
9	2	always be set to zero.
11	4	offset from the beginning of the file to the bitmap data.
The bitmap information header		
15	4	size of the bitmap information header structure, in bytes.
19	4	width of the image, in pixels.
23	4	height of the image, in pixels.
27	2	set to zero.
29	2	number of bits per pixel.
31	4	compression, usually set to zero, 0 = no compression.
35	4	size of the image data, in bytes.
39	4	horizontal pixels per metre, usually set to zero.
43	4	vertical pixels per metre, usually set to zero.
47	4	number of colours used in the bitmap.
51	4	number of colours that are 'important' for the bitmap, if set to zero, all colours are important.

```
42 4D 36 96 10 00 00 00 00 00 76 00 00 00 28 00   BM6-......v...(.
00 00 C4 04 00 00 F0 06 00 00 01 00 04 00 00 00   ..Ä...ð.........
00 00 C0 95 10 00 43 2E 00 00 43 2E 00 00 10 00   ..À•..C...C.....
00 00 10 00 00 00 47 50 50 00 5A 68 63 00 5B 6B   ......GPP.Zhc.[k
70 00 67 6E 6F 00 62 73 77 00 63 78 82 00 72 7B   p.gno.bsw.cx,.r{
7E 00 76 87 8D 00 81 94 9D 00 8A A3 AD 00 9C AC   ~.v‡□.□"□.Š£-.œ¬
B3 00 A4 B8 C0 00 9F BD CA 00 AC BF CC 00 B2 CD   ³.¤¸À.Ÿ½Ê.¬¿Ì.²Í
D4 00 C1 DE EC 00 66 66 46 44 66 77 64 66 66 66   Ô.ÁÞì.ffFDfwdfff
66 66 66 66 66 66 54 66 66 66 66 66 66 44 44 43   ffffffTffffffDDC
33 22 33 33 33 33 34 44 43 33 33 22 23 33 33 33   3"33334DC33"#333
33 33 34 44 66 44 44 44 44 44 44 44 66 44 44   334DfDDDDDDDfDD
44 44 46 66 66 66 66 66 66 66 44 44 44 44 44 44   DDFfffffffDDDDDD
```

In decimal, what is the:

	Hex values, little endian	Hex values, big endian	Decimal value
Image file size			
Image width			
Image height			
Number of bits per pixel			
Max. number of colours possible	n/a	n/a	

Worksheet   Unit 1.21 Storage of data, bitmaps

The hex dump below is the top of a bitmapped file. Look at the values in the file header and fill in the correct values in the table.

Bitmap image file format details

Start position	Size in bytes	Meaning or use
The bitmap file header		
1	2	set to 'BM' to show this is a .bmp-file.
3	4	size of the file in bytes.
7	2	always be set to zero.
9	2	always be set to zero.
11	4	offset from the beginning of the file to the bitmap data.
The bitmap information header		
15	4	size of the bitmap information header structure, in bytes.
19	4	width of the image, in pixels.
23	4	height of the image, in pixels.
27	2	set to zero.
29	2	number of bits per pixel.
31	4	compression, usually set to zero, 0 = no compression.
35	4	size of the image data, in bytes.
39	4	horizontal pixels per metre, usually set to zero.
43	4	vertical pixels per metre, usually set to zero.
47	4	number of colours used in the bitmap.
51	4	number of colours that are 'important' for the bitmap, if set to zero, all colours are important.

```
42 4D 62 D6 01 00 00 00 00 00 36 04 00 00 28 00   BMbÖ......6...(.
00 00 52 01 00 00 5F 01 00 00 01 00 08 00 00 00   ..R..._..........
00 00 2C D2 01 00 23 2E 00 00 23 2E 00 00 00 01   ..,Ò..#...#.....
00 00 00 01 00 00 00 00 00 00 01 01 01 00 02 02   ................
02 00 03 03 03 00 04 04 04 00 05 05 05 00 06 06   ................
06 00 07 07 07 00 08 08 08 00 09 09 09 00 0A 0A   ...............
0A 00 0B 0B 0B 00 0C 0C 0C 00 0D 0D 0D 00 0E 0E   ...............
0E 00 0F 0F 0F 00 10 10 10 00 11 11 11 00 12 12   ...............
12 00 13 13 13 00 14 14 14 00 15 15 15 00 16 16   ...............
16 00 17 17 17 00 18 18 18 00 19 19 19 00 1A 1A   ...............
1A 00 1B 1B 1B 00 1C 1C 1C 00 1D 1D 1D 00 1E 1E   ...............
1E 00 1F 1F 1F 00 20 20 20 00 21 21 21 00 22 22   ...... .!!!.""
```

In decimal, what is the:

	Hex values, little endian	Hex values, big endian	Decimal value
Image file size			
Image width			
Image height			
Number of bits per pixel			
Max. number of colours possible	n/a	n/a	

Worksheet   Unit 1.22 Storage of data, bitmaps

The hex dump below is the top of a bitmapped file. Look at the values in the file header and fill in the correct values in the table.

Bitmap image file format details

Start position	Size in bytes	Meaning or use
The bitmap file header		
1	2	set to 'BM' to show this is a .bmp-file.
3	4	size of the file in bytes.
7	2	always be set to zero.
9	2	always be set to zero.
11	4	offset from the beginning of the file to the bitmap data.
The bitmap information header		
15	4	size of the bitmap information header structure, in bytes.
19	4	width of the image, in pixels.
23	4	height of the image, in pixels.
27	2	set to zero.
29	2	number of bits per pixel.
31	4	compression, usually set to zero, 0 = no compression.
35	4	size of the image data, in bytes.
39	4	horizontal pixels per metre, usually set to zero.
43	4	vertical pixels per metre, usually set to zero.
47	4	number of colours used in the bitmap.
51	4	number of colours that are 'important' for the bitmap, if set to zero, all colours are important.

```
42  4D  36  50  1D  00  00  00  00  00  36  04  00  00  28  00    BM6P......6...(.
00  00  40  06  00  00  B0  04  00  00  01  00  08  00  00  00    ..@...°.........
00  00  00  4C  1D  00  13  0B  00  00  13  0B  00  00  00  01    ...L............
00  00  00  01  00  00  00  00  00  00  01  01  01  00  02  02    ................
02  00  03  03  03  00  04  04  04  00  05  05  05  00  06  06    ................
06  00  07  07  07  00  08  08  08  00  09  09  09  00  0A  0A    ................
0A  00  0B  0B  0B  00  0C  0C  0C  00  0D  0D  0D  00  0E  0E    ................
0E  00  0F  0F  0F  00  10  10  10  00  11  11  11  00  12  12    ................
12  00  13  13  13  00  14  14  14  00  15  15  15  00  16  16    ................
16  00  17  17  17  00  18  18  18  00  19  19  19  00  1A  1A    ................
1A  00  1B  1B  1B  00  1C  1C  1C  00  1D  1D  1D  00  1E  1E    ................
1E  00  1F  1F  1F  00  20  20  20  00  21  21  21  00  22  22    ......   .!!!."″
```

In decimal, what is the:

	Hex values, little endian	Hex values, big endian	Decimal value
Image file size			
Image width			
Image height			
Number of bits per pixel			
Max. number of colours possible	n/a	n/a	

Worksheet   Unit 1.23 Storage of data, bitmaps

The hex dump below is the top of a bitmapped file. Look at the values in the file header and fill in the correct values in the table.

Bitmap image file format details

Start position	Size in bytes	Meaning or use
The bitmap file header		
1	2	set to 'BM' to show this is a .bmp-file.
3	4	size of the file in bytes.
7	2	always be set to zero.
9	2	always be set to zero.
11	4	offset from the beginning of the file to the bitmap data.
The bitmap information header		
15	4	size of the bitmap information header structure, in bytes.
19	4	width of the image, in pixels.
23	4	height of the image, in pixels.
27	2	set to zero.
29	2	number of bits per pixel.
31	4	compression, usually set to zero, 0 = no compression.
35	4	size of the image data, in bytes.
39	4	horizontal pixels per metre, usually set to zero.
43	4	vertical pixels per metre, usually set to zero.
47	4	number of colours used in the bitmap.
51	4	number of colours that are 'important' for the bitmap, if set to zero, all colours are important.

```
42 4D C2 86 20 00 00 00 00 00 36 00 00 00 28 00   BMÂ† .....6...(.
00 00 CB 03 00 00 DB 02 00 00 01 00 18 00 00 00   ..Ë...Û.........
00 00 8C 86 20 00 30 5C 00 00 30 5C 00 00 00 00   ..Œ† .0\..0\....
00 00 00 00 00 00 31 45 4A 31 45 4A 31 45 4A 31   ......1EJ1EJ1EJ1
45 4A 31 45 4A 39 49 4A 39 4D 52 39 4D 52 39 4D   EJ1EJ9IJ9MR9MR9M
52 39 4D 52 39 4D 52 39 4D 52 39 4D 52 39 49 52   R9MR9MR9MR9MR9IR
39 49 52 39 49 4A 31 49 4A 31 49 52 31 49 52 31   9IR9IJ1IJ1IR1IR1
49 52 39 4D 52 39 55 5A 42 59 5A 42 59 5A 39 55   IR9MR9UZBYZBYZ9U
5A 42 59 5A 42 59 63 39 55 5A 39 55 5A 39 55 5A   ZBYZBYc9UZ9UZ9UZ
42 55 5A 42 51 5A 39 51 5A 42 55 5A 4A 5D 63 52   BUZBQZ9QZBUZJ]cR
69 73 5A 71 7B 63 79 84 63 79 84 63 79 84 5A 75   isZq{cy„cy„cy„Zu
7B 5A 75 7B 5A 75 7B 5A 71 7B 5A 71 7B 5A 6D 73   {Zu{Zu{Zq{Zq{Zms
```

In decimal, what is the:

	Hex values, little endian	Hex values, big endian	Decimal value
Image file size			
Image width			
Image height			
Number of bits per pixel			
Max. number of colours possible	n/a	n/a	

Answers to worksheets

Worksheet   Unit 1.1 Stored numbers, integers

Signed	Unsigned
183	183
89	89
23 912	23 912
−349	33 117
9481	9481
−7399	40 167

Worksheet   Unit 1.2 Stored numbers, integers

Signed	Unsigned
28 550	28 550
29 374	29 374
15 316	15 316
29 187	29 187
26 692	26 692
4197	4197

Worksheet   Unit 1.3 Stored numbers, integers

Signed	Unsigned
5632	5632
14 286	14 286
10 624	10 624
8728	8728
13 022	13 022
26 001	26 001

Worksheet   Unit 1.4 Stored numbers, integers

Signed	Unsigned
28 748	28 748
11 082	11 082
26 044	26 044
18 740	18 740
26 218	26 218
8733	8733

Worksheet   Unit 1.5 Stored numbers, integers

Signed	Unsigned
22 807	22 807
25 774	25 774
10 294	10 294
7789	7789
3632	3632
28 705	28 705

Worksheet   Unit 1.6 Stored numbers, integers

Signed	Unsigned
−29 241	62 009
−25 401	58 169
15 219	15 316
15 408	29 187
−714	33 482
−12 406	45 174

Worksheet   Unit 1.7 Stored numbers, integers

Signed	Unsigned
−13 334	46 102
6693	6693
1932	1932
−14 691	47 459
19 677	19 677
−17 121	49 889

Worksheet   Unit 1.8 Stored numbers, integers

Signed	Unsigned
29 068	29 068
19 032	19 032
21 132	21 132
−29 789	62 557
23 544	23 544
−14 835	47 603

Worksheet   Unit 1.9 Stored numbers, integers

Signed	Unsigned
10 347	10 347
25 855	25 855
4907	4907
−7140	39 908
−17 751	50 519
42	42

Worksheet   Unit 1.10 Stored numbers, integers

Signed	Unsigned
−29 568	62 336
24 323	24 323
5883	5883
1823	1823
28 178	28 178
684	684

Worksheet   Unit 1.11 Floating point fractional numbers

Possible functions to use in the spreadsheet

A	B	C	D
1	0.1	=	=CONCATENATE(B65,B66)
2			
3	=B1	=B3*2	=INT(C3)
4	=C3-D3	=B4*2	=INT(C4)
5	=C4-D4	=B5*2	=INT(C5)
6	=C5-D5	=B6*2	=INT(C6)
7	=C6-D6	=B7*2	=INT(C7)
8	=C7-D7	=B8*2	=INT(C8)
	Rows deleted to save space		
57	=C56-D56	=B57*2	=INT(C57)
58	=C57-D57	=B58*2	=INT(C58)
59	=C58-D58	=B59*2	=INT(C59)
60	=C59-D59	=B60*2	=INT(C60)
61	=C60-D60	=B61*2	=INT(C61)
62	=C61-D61	=B62*2	=INT(C62)
63			
64			
65	=CONCATENATE(".",D3,D4,D5,D6,D7,D8,D9,D10,D11,D12,D13,D14,D15,D16,D17,D18,D19, D20,D21,D22,D23,D24,D25,D26,D27,D28,D29,D30,D31,D32)		
66	=CONCATENATE(D33,D34,D35,D36,D37,D38,D39,D40,D41,D42,D43,D44,D45,D46,D47,D48, D49,D50,D51,D52,D53,D54,D55,D56,D57,D58,D59,D60,D61)		

Concatenate means 'join together'. Microsoft Excel can only join up to 30 items with the concatenate function so this spreadsheet uses two concatenate functions then joins the results.

Decimal	60 place binary	48-bit accurate?
0.1	.000110011001100110011001100110011001100110011001100110011010000	No
0.25678	.010000011011110001010101100001100100010001010010010000000000	No
0.0625	.000100000000000000000000000000000000000000000000000000000000	Yes
0.251220703	.010000000101000000000000000000000000000000000000000000000000	Yes
0.141592654	.001001000011111011010101000100010000101101000110000000000000	Yes

Worksheet Unit 1.12 Number bases

Decimal	Hex	Octal	Binary
117212	1C9DC	344734	0001110010011101
1431	597	2627	010110010111
1044	414	2024	010000010100
4	4	4	0100
91509	16575	262565	0001011001010111
1248	4E0	2340	010011100000
69	45	105	01000101
100875	18A0B	305013	0001100010100000
340	154	524	000101010100
66182	10286	201206	0001000000101000
8712	2208	21010	0010001000001000
10815	2A3F	25077	0010101000111111

Worksheet Unit 1.13 Number bases

Decimal	Hex	Octal	Binary
70149	11205	211005	0001000100100000
4252	109C	10234	0001000010011100
92663	169F7	264767	0001011010011111
47	2F	57	00101111
10229	27F5	23765	0010011111110101
232	E8	350	11101000
589	24D	1115	001001001101
26805	68B5	64265	0110100010110101
221	DD	335	11011101
3160	C58	6130	110001011000
8813	226D	21155	0010001001101101
445	1BD	675	000110111101

Worksheet Unit 1.14 Number bases

Decimal	Hex	Octal	Binary
10272	2820	24040	0010100000100000
4599	11F7	10767	0001000111110111
10308	2844	24104	0010100001000100
8246	2036	20066	0010000000110110
377	179	571	000101111001
2947	B83	5603	101110000011
6443	192B	14453	0001100100101011
11389	2C7D	26175	0010110001111101
9148	23BC	21674	0010001110111100
10332	285C	24134	0010100001011100
127139	1F0A3	370243	0001111100001010
3479	D97	6627	110110010111

Worksheet   Unit 1.15 Number bases

Decimal	Hex	Octal	Binary
18 318	478E	43616	0100011110001110
7948	1F0C	17414	0001111100001100
8326	2086	20206	0010000010000110
6443	192B	14453	0001100100101011
1025	401	2001	010000000001
93 208	16C18	266030	0001011011000001
98 757	181C5	300705	0001100000011100
4734	127E	11176	0001001001111110
55 845	DA25	155045	1101101000100101
11 453	2CBD	26275	0010110010111101
120 084	1D514	352424	0001110101010001
210	D2	322	11010010

Worksheet   Unit 1.16 Number bases

Decimal	Hex	Octal	Binary
350	15E	536	000101011110
663	297	1227	001010010111
510	1FE	776	000111111110
1706	6AA	3252	011010101010
8742	2226	21046	0010001000100110
72 928	11CE0	216340	0001000111001110
8702	21FE	20776	0010000111111110
8458	210A	20412	0010000100001010
12 469	30B5	30265	0011000010110101
4272	10B0	10260	0001000010110000
77	4D	115	01001101
5474	1562	12542	0001010101100010

Worksheet   Unit 1.17 Number bases

Decimal	Hex	Octal	Binary
12 423	3087	30207	0011000010000111
8683	21EB	20753	0010000111101011
74 677	123B5	221665	0001001000111011
112 061	1B5BD	332675	00011011010110111101
359	167	547	000101100111
233	E9	351	11101001
2973	B9D	5635	101110011101
123 366	1E1E6	360746	0001111000011110
29	1D	35	00011101
10 778	2A1A	25032	0010101000011010
786	312	1422	001100010010
10 167	27B7	23667	0010011110110111

Worksheet   Unit 1.18 Storage of data, bitmaps

	Hex values, little endian	Hex values, big endian	Decimal value
Image file size	36 1B B7 00	B7 1B 36	12 000 054
Image width	D0 07 00 00	7D0	2000
Image height	D0 07 00 00	7D0	2000
Number of bits per pixel	18 00	18	24
Max. number of colours possible	n/a	n/a	16.7 million

Worksheet   Unit 1.19 Storage of data, bitmaps

	Hex values, little endian	Hex values, big endian	Decimal value
Image file size	8E 37 38 00	38378E	3 684 238
Image width	22 09 00 00	922	2338
Image height	26 06 00 00	626	1574
Number of bits per pixel	08 00	8	8
Max. number of colours possible	n/a	n/a	256

Worksheet   Unit 1.20 Storage of data, bitmaps

	Hex values, little endian	Hex values, big endian	Decimal value
Image file size	36 96 10 00	109636	1 087 030
Image width	C4 04 00 00	4C4	1220
Image height	F0 06 00 00	6F0	1776
Number of bits per pixel	04 00	4	4
Max. number of colours possible	n/a	n/a	16

Worksheet   Unit 1.21 Storage of data, bitmaps

	Hex values, little endian	Hex values, big endian	Decimal value
Image file size	62 D6 01 00	1D662	120 418
Image width	52 01 00 00	152	338
Image height	5F 01 00 00	15F	351
Number of bits per pixel	08 00	8	8
Max. number of colours possible	n/a	n/a	256

Worksheet   Unit 1.22 Storage of data, bitmaps

	Hex values, little endian	Hex values, big endian	Decimal value
Image file size	36 50 1D 00	1D5036	1 921 078
Image width	40 06 00 00	640	1600
Image height	B0 04 00 00	4B0	1200
Number of bits per pixel	08 00	8	8
Max. number of colours possible	n/a	n/a	256

Worksheet   Unit 1.23 Storage of data, bitmaps

	Hex values, little endian	Hex values, big endian	Decimal value
Image file size	C2 86 20 00	2086C2	2 131 650
Image width	CB 03 00 00	3CB	971
Image height	DB 02 00 00	2DB	731
Number of bits per pixel	18 00	18	24
Max. number of colours possible	n/a	n/a	16.7 million

1.6 Assignment

Hierarchy of design

In computing in general, it is very important that students understand the hierarchical nature of systems and their design. Gone are the days when monolithic designs were commonplace; modern products benefit from the ideas of layered, modularized designs. This has led to the much improved ability to modify parts, extend their functionality and flexibility and to improve reliability.

This idea applies equally to networks (ISO7 layer model etc.) and to the design of hardware.

The object of this assignment is to foster the concepts of the hierarchy of design.

Your assignment is to provide an example of the hierarchy of design leaving the computer hardware and software elements left in place.

Example repeated from the book for convenience.

of bricks and towns	of computer hardware	of computer software
Bricks. Study what a brick is made of, how it is made, how strong it is, what will it cost	1s and 0s, simple digital circuits and how logical arithmetic can be performed with a circuit	Boolean logic
Walls. Study how to mix cement, how to lay bricks to make a wall, how strong is a wall	How a sequence of logical operations can be achieved with a circuit, how to add, subtract, perform logical AND and OR operations etc.	How to perform arithmetic with simple numbers
How to make several walls into a building with spaces for windows and doors etc. How to build a roof	How to store many logical instructions and feed them in sequence to a circuit that can execute them	How to perform arithmetic with multiple digit numbers
How to install all the services a building needs, water, electricity, gas, heating etc. and to move in the carpets, furniture etc.	How to accept human inputs by devices such as a keyboard and to display outputs using devices like a colour monitor	How to handle data such as text and to edit it, i.e. move a sentence within a paragraph
How to build a row of houses, provide street lighting, public access etc.	How to provide a complete set of devices such as a mouse, keyboard, printer, CPU etc. and to make them all connect correctly	How to present a complete set of facilities in a word processor
How to plan a town, provide libraries, shops, hospital, bus station etc.	A complete PC	How to control the entire machine: the operating system

Example answer

of car design	of computer hardware	of computer software
How petrol burns, its chemistry and physical properties. The metallurgy of engine parts	1s and 0s, simple digital circuits and how logical arithmetic can be performed with a circuit	Boolean logic
Design of wheels, gears, pistons etc., their individual characteristics and limitations	How a sequence of logical operations can be achieved with a circuit, how to add, subtract, perform logical AND and OR operations etc.	How to perform arithmetic with simple numbers
How to make an engine from individual components, how much power it produces, how fuel efficient it is	How to store many logical instructions and feed them in sequence to a circuit that can execute them	How to perform arithmetic with multiple digit numbers
How to assemble the engine, gearbox, wheels, seats, body shell and all the other components into a complete car	How to accept human inputs by devices such as a keyboard and to display outputs using devices like a colour monitor	How to handle data such as text and to edit it, i.e. move a sentence within a paragraph
How the complete car performs, how acceptable it is as a product to customers in terms of comfort, performance, reliability etc.	How to provide a complete set of devices such as a mouse, keyboard, printer, CPU etc. and to make them all connect correctly	How to present a complete set of facilities in a word processor
The complete design and manufacturing details for mass production of the car	A complete PC	How to control the entire machine: the operating system

Layered architecture

In computing in general, it is very important that students understand the hierarchical nature of systems and their design. Gone are the days when monolithic designs were commonplace, modern products benefit from the ideas of layered, modularized designs. This has led to the much improved ability to modify parts, extend their functionality and flexibility and to improve reliability.

This idea applies equally to networks (ISO7 layer model etc.) and to the design of hardware.

The object of this assignment is to foster the concepts of the hierarchy of design.

Your assignment is to present a conversation between students and a teacher in the seven layers of the ISO model. Imagine a group of 12 students and a teacher are seated in a room and are talking as a group on one subject. Select each aspect of communication and assign that aspect to one of the seven layers. Remember that it is communication that is the aim and that you cannot stretch the analogy between speech and networks too far, mainly because humans are far more complex than machines.

OSI/ISO 7 layer model	Spoken communication analogue
Layer 7 The application layer	
Layer 6 The presentation layer	
Layer 5 The session layer	
Layer 4 The transport layer	
Layer 3 The network layer	
Layer 2 The data link layer	
Layer 1 The physical layer	

Example answer

OSI/ISO 7 layer model	Spoken communication analogue
Layer 7 The application layer	Speak about the subject you wish to communicate. 'Tell me about the weather today?', 'Who will win the election?', 'How are you feeling today?'
Layer 6 The presentation layer	Use of jargon or plain language (does everyone communicating know the meaning of the jargon?). Speak in French or English or any other human language
Layer 5 The session layer	Start or end the conversation, possibly with 'Good morning' or 'May I speak now?' 'That's all for today, thanks and goodbye'. Not concerned with subject only the establishment of a communication session. Most people would call this 'starting (or ending) a conversation'.
Layer 4 The transport layer	If the hearer has not heard, how do you know they have not heard? They may send back a message: 'Pardon?' or 'I didn't quite catch what you said', i.e. the message has arrived but has not been completely received. This is a form of error checking
Layer 3 The network layer	Look or point at the person to whom you wish to speak; engage eye contact. Establish a path to someone you wish to speak to
Layer 2 The data link layer	Syllables of speech, separate sounds, not speaking if someone else is speaking (collision detection)
Layer 1 The physical layer	The sound travelling through the air or over a telephone line, megaphone or radio link

Systems analysis

Rationale

By way of an introduction students will learn:

- why software development techniques are required in the analysis/design process
- how to apply software analysis techniques for a given 'real world' specification based on an appropriate lifecycle model
- about different approaches to analysis and make comparisons about different models that are used within the software engineering environment today
- about the techniques involved in 'fact finding' for a systems investigation and how to record/present follow-up documentation for a prospective customer
- about basic data modelling techniques, including functional modelling development and graphical representations (for example, entity-relationship diagrams) and to implement such designs into a simple rational database system.

The unit provides a foundation of systems analysis and design by covering requirements analysis skills for both commercial and technical applications. It also provides an underlying structure by developing the key skills required within the field of software engineering.

Students will need to be exposed to a variety of learning activities and assignments covering both theoretical and practical aspects of the unit. Students need to be able to produce documentation to a professional standard which would need to include graphical representations. Students should be able to analyse the requirements of a proposed system, document the outcomes and communicate the results to a prospective customer. They should be aware of information systems in use within industry and be able to design and implement a simple relational database system which could be based on documented analysis developed under systems investigation.

This unit is designed to integrate analysis techniques with practical applications in order for students to use information technology to document, analyse, design and implement systems being investigated.

2.1 Systems analysis lifecycle

Additional lifecycle exercise

The traditional software lifecycle has been around for many years and is now coming under increased criticism by many developers. You are to investigate what are the main criticisms of using this technique to develop a modern software solution and how the cycle can be modified to meet these needs.

Suggested solution

- The lifecycle is always shown to be linear in structure. The process follows directly through from one stage to the next until the maintenance stage. Each new stage starting on the completion of the previous stage. In most software development applications this process is not satisfactory, as it needs to be iterative. There needs to be a looping structure built into the development process so that the team can cycle around in order to build on and further develop ideas/concepts fundamental to its success. The figure below outlines a problem that occurs in the implementation phase and is reflected back up the detailed design phase in order to modify the problem. This iteration may occur several times and may have to reflect back up the lifecycle further in order to rectify the problem or modification requirement.

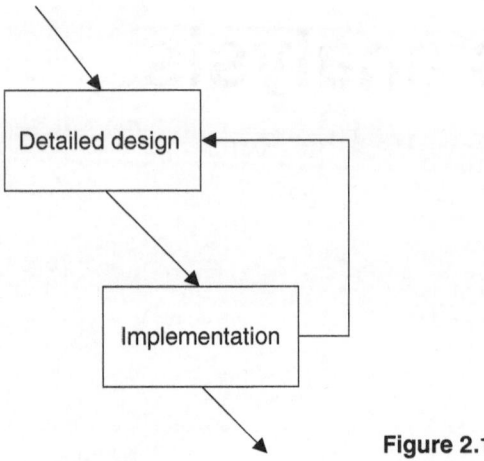

Figure 2.1

- During the development the customer may decide that the original requirements did not cover all aspects and that they should now be modified. This may mean that the team have to work back to the beginning of the lifecycle to incorporate the new requirements. It has been found that many initial specifications do not fully specify all expected requirements and a level of uncertainty exists that can only be handled by an iterative process.
- The traditional lifecycle does not allow the customer to see the working software until the end of the process. If a problem is detected at this stage it could have serious consequences for the software developer. As we shall see next this is one of the advantages of using a prototype model.

The waterfall definition

The classic software lifecycle is often represented as a simple *prescriptive* waterfall software phase model, where software evolution proceeds through an orderly sequence of transitions from one phase to the next in order (Royce, 1970). Such models resemble finite state machine descriptions of software evolution. However, these models have perhaps been most useful in helping to structure, staff, and manage large software development projects in complex organizational settings, which was one of the primary purposes (Royce, 1970; Boehm, 1976). Alternatively, these classic models have been widely characterized as both poor *descriptive* and *prescriptive* models of how software development 'in-the-small' or 'in-the-large' can or should occur.

prescriptive

A prescriptive model prescribes how a new software system should be developed. Prescriptive models are used as guidelines or frameworks to organize and structure how software development activities should be performed, and in what order. Typically, it is easier and more common to articulate a prescriptive lifecycle model for how systems should be developed. This is possible since most such models are intuitive or well reasoned. This means that many idiosyncratic details that describe how a software system is built in practice can be ignored, generalized, or deferred for later consideration. This, of course, should raise concern for the relative validity and robustness of such lifecycle models when developing different kinds of applications systems, in different kinds of development settings, using different programming languages, with differentially skilled staff etc. However, prescriptive models are also used to package the development tasks and techniques for using a given set of software engineering tools or environment during a development project.

descriptive

A descriptive model describes the history of how a particular software system was developed. Descriptive models may be used as the basis for understanding and improving the software development process, or for building empirically grounded prescriptive models. Descriptive lifecycle models characterize how particular software systems are actually developed in specific settings. As such, they are less common and more difficult to articulate for an obvious reason: one must observe or collect data throughout the lifecycle of a software system, a period of elapsed time often measured in years. Also, descriptive models are specific for the system observed and only generalized through systematic comparative analysis. Therefore, it suggests

the prescriptive software lifecycle models will dominate attention until a sufficient base of observational data is available to articulate empirically grounded descriptive lifecycle models.
(Reference IEEE 1074/1995 and Scacchi, 2001)

Further information on rapid software prototyping

Background

A rapid software prototype is a dynamic visual model providing a means of communication for both the customer and developer. It is more effective than textual documents or static models for displaying limited working applications of the proposed system. A prototype contains the following characteristics:

- A functional model can be displayed after a small amount of work
- Provides a flexible base as modifications can easily be implemented
- Provides the end user with an overview of the proposed system that shows the key components before full implementation
- Prototype activities are completed quickly occurring small costs and time constraints
- A prototype does not always provide a true representation of the complete system.

Benefits of the prototype model:

- Any misunderstandings between the software developers and the end users are highlighted early in the development process
- Missing functional components may be detected
- Components that are difficult to use or confusing may be identified
- A limited working system is available for end users, developers and management to see
- It serves as a base for developing the complete system specification.

From Pressman

> The prototyping principle, as presented by Pressman, focuses on building a sequence of prototypes until the requirements are understood, thereby assisting the customer or developer. The collection of requirements is the first step in this process. The next step is to develop a quick design. This design is only focused on the parts of the system that the user is concerned with and is turned into a prototype. Next, the user tests the prototype and a new discussion concerning the requirements for the system takes place. This cycle of talking to the customer, building a prototype, and then testing it is continued until all parties involved feel that the requirements are complete.

Rapid software prototyping is basically a systems analysis technique that provides a base for an accurate and complete set of functional requirements for the proposed system. It overcomes the problem that the traditional lifecycle model has in that the user can view a physical model early in the development process and not at the final testing phase. But it does have its drawbacks, such as adequate prototyping skills, mistaken concepts concerning definitions, objectives and functional requirements which can lead to 'short cuts' within the full working application; premature delivery of the software product and its associated documentation; and problems with standardization and uniformity especially if fourth generation languages are used as generic tools during development.
See:

http://www.shu.ac.uk/schools/cms/rapid.software.prototyping/

The prototyping lifecycle – additional exercise

Like the traditional lifecycle prototyping models have their own cycle of activities. Develop a suitable model that can be used regardless of the prototyping process being used.

Suggested solution

Product development will normally move along four lines during the completion of the work:

- There are the initial ideas which finally turn into the completed project
- There are the diagrammatic analysis and design tools that turn into the final code
- You move from a low technology base (through meetings and customer interaction) to a high technology base for the final software and hardware requirements

- We do not know about the final performance at the inset of a project, we just have a general appearance as to what is required.

Prototyping must be an iterative process and must support rapid development and modification of the prototype models. The main steps in this iterative process are:

- *Plan* – understand the functional requirements and the needs of the end users and how these needs are to be incorporated into the design process
- *Implement* – develop the prototype model to implement the concepts developed in the planning activity
- *Measure* – here you can see if the prototype matches the needs of the customer. The user needs to check/test the model to see if there are any problems and if any additional concepts are needed. The developer will be looking for a quality product but has to be aware of the time constraints for delivering the product.
- *Learn* – this phase allows the developer to analyse the work carried out so far, learn from any new development/mistakes and see which parts of the prototyping model are doing well or not.

The outcome should be a rapid prototyping model that produces the following:

- An early insight into the product
- The end user can test the product early in its cycle
- The end user can have enhanced feedback during development
- The requirements can be clearly defined and modified if required
- Fast development process within the prototype models.

See the role of prototyping in software development at:

http://www.swe.uni-linz.ac.at/publications/abstract/TR-SE-94.02.html

Note: This is a useful PDF document and includes prototyping for object software development.

2.2 Systems analysis tools and techniques

Note: These exercises have been completed within the Select Yourdon CASE tool environment. Full details about the techniques used are contained in their user documentation. In order for the diagrams to be fully syntactically correct additional information needs to be added, for example Store Balancing and Group Node BNF structures. Once all the refinements have been completed they need to be checked for consistency to ensure all the 'parent–child balancing' links are correct.

Solution to Exercise 2.2.1 – Bess and Bailey Dog Kennels
(i) Context diagram

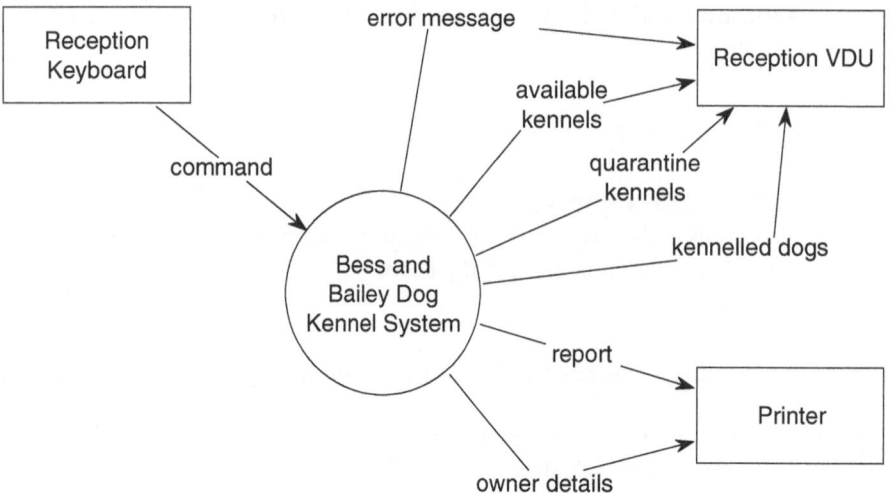

Figure 2.2 *Bess and Bailey Kennel System: Context diagram*

Check results for the context diagram

```
Project: C:\SELECT\SYSTEM\BBDC01\
Title: Bess and Bailey Dog Kennels
Date: 13-Dec-2000 Time: 11:45

Checking DOG1.DAT

No Errors detected, No Warnings given.

----- End of report -----
```

(ii) First refinement

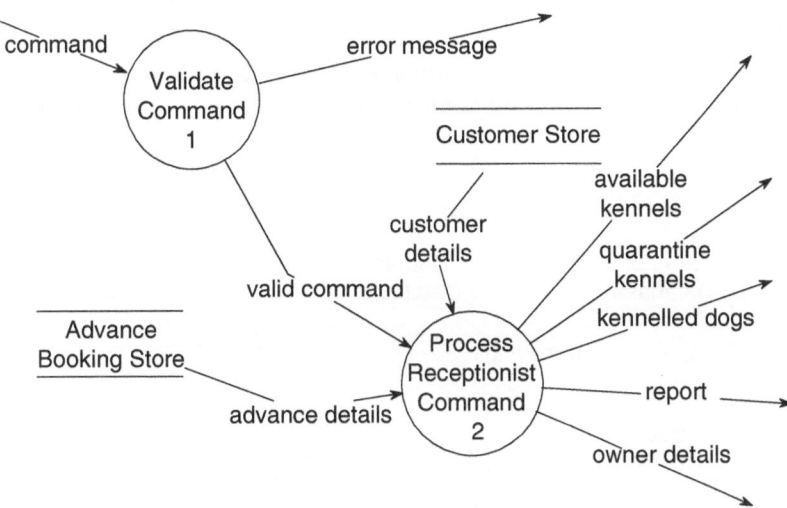

Figure 2.3 *Bess and Bailey Kennel System: first refinement*

Test of the first refinement:

```
Project: C:\SELECT\SYSTEM\BBDC01\
Title: Bess and Bailey Dog Kennels
Date: 13-Dec-2000 Time: 12:3

Checking DOG2.DAT

No Errors detected, No Warnings given.

----- End of report -----
```

Store balancing requirements (only one input from the stores, but they still need to be added to the BNF clause):

```
Project: C:\SELECT\SYSTEM\BBDC01\
Title: Bess and Bailey Dog Kennels
Date: 13-Dec-2000 Time: 12:3

Name: Customer Store
Type: Store
Bnf: customer details
This item is used on the following diagrams:
   DOG2.DAT  Bess and Bailey Dog Kennel
System
Last changed: DEFAULT 13-Dec-2000 12:22:25

----- End of report -----
```

Project: C:\SELECT\SYSTEM\BBDC01\
Title: Bess and Bailey Dog Kennels
Date: 13-Dec-2000 Time: 12:4

Name: **Advance Booking Store**
Type: Store
Bnf: advance details
This item is used on the following diagrams:
    DOG2.DAT Bess and Bailey Dog Kennel
System
Last changed: DEFAULT 13-Dec-2000 12:21:19

----- End of report -----

Note: The system will need a central database that will contain information on the kennels, which are set aside for quarantine, which have dogs in (and who they are) and which are free to use. This database will be the hub of most of the kennel commands and booking requirements and in respect to this it will be introduced in the second refinement.

Second refinement

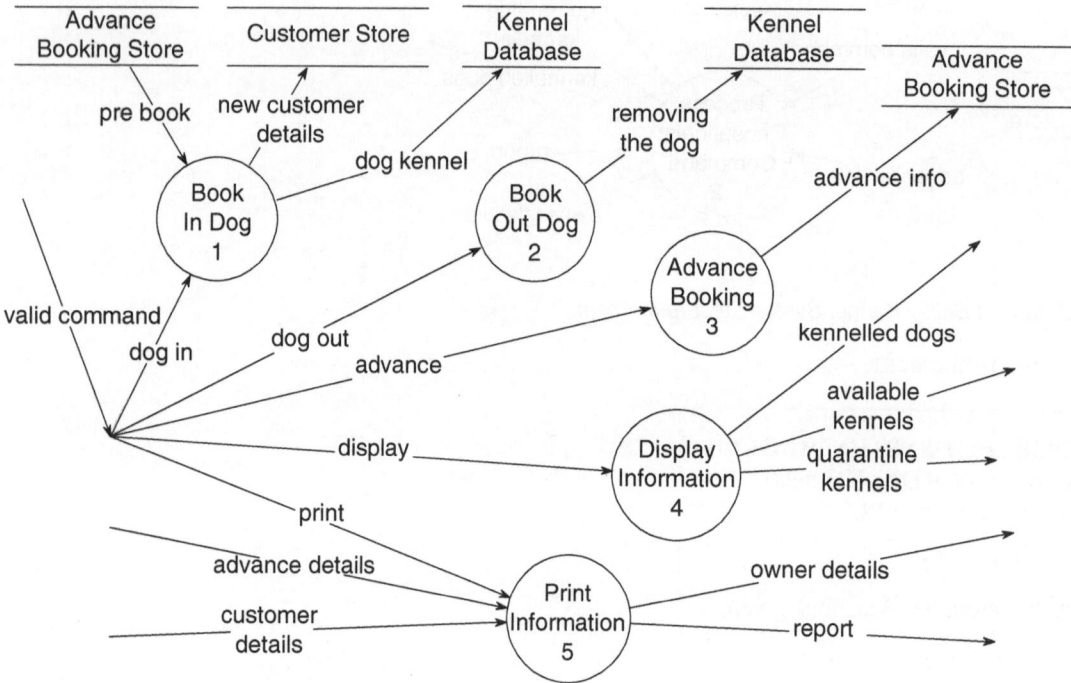

Figure 2.4 *Process receptionist command: second refinement*

Check on the Process Receptionist Command

Project: C:\SELECT\SYSTEM\BBDC01\
Title: Bess and Bailey Dog Kennels
Date: 19-Dec-2000 Time: 18:3

Checking DOG3.DAT

No Errors detected, No Warnings given.

----- End of report -----

Process 4 (Display Information) and Process 5 (Print Information) require further refinements. Processes 1, 2 and 3 require no further refinements so their Process Specifications (PSPECs) are shown below. Remember you can continue to decompose the diagrams until a process is detailed enough to become a Process Specification (PSPEC).

(iii) Process Specification for the Book In Dog process

@IN = dog in
@IN = pre book
@OUT = dog kennel
@OUT = new customer details

@PSPEC 0.2.1 **Book In Dog**

On receiving the dog in command do:
check the Advanced Booking Store to ascertain the pre book
requirements for the customer (if any)
and
update the Kennel Database by sending the new dog kennel data
and
update the Customer Store by sending the new customer details

@

Process Specification for the Display Available Kennels process

@IN = dog out
@OUT = removing the dog

@PSPEC 0.2.2 **Book Out Dog**

On receiving the dog out command do:
update the Kennel Database by removing the dog
from the kennels it had occupied

@

Process Specification for the Display Available Kennels process

@IN = advance
@OUT = advance info

@PSPEC 0.2.3 **Advance Booking**

On receiving the advance command do:
send the customer advance info to the Advance Booking
Store

@

Note: All the PSPECs should be checked, full details of these will be shown in the final consistency check.

Group Node BNF requirements for the **valid command** data flow

Project: C:\SELECT\SYSTEM\BBDC01\
Title: Bess and Bailey Dog Kennels
Date: 20-Dec-2000 Time: 9:2

Name: **valid command**
Type: Discrete flow
Bnf: [dog in | dog out | advance | display | print]
This item is used on the following diagrams:
DOG2.DAT   Bess and Bailey Dog Kennel System
DOG3.DAT   Process Receptionist Command
Last changed: DEFAULT   19-Dec-2000 19:06:01

----- End of report -----

The BNF statement shows an 'or' situation and the receptionist can enter one command or another and not all of them at the same time.

Store balancing – Modification to the Advance Booking Store and Customer Store

```
Project: C:\SELECT\SYSTEM\BBDC01\
Title: Bess and Bailey Dog Kennels
Date: 20-Dec-2000 Time: 9:3


Name: Advance Booking Store
Type: Store
Bnf: [ advance details | pre book | advance info ]
This item is used on the following diagrams:
        DOG2.DAT    Bess and Bailey Dog Kennel System
        DOG3.DAT    Process Receptionist Command
Last changed: DEFAULT    13-Dec-2000 13:15:40

----- End of report -----
```

```
Project: C:\SELECT\SYSTEM\BBDC01\
Title: Bess and Bailey Dog Kennels
Date: 20-Dec-2000 Time: 9:3


Name: Customer Store
Type: Store
Bnf: [ customer details | new customer details ]
This item is used on the following diagrams:
        DOG2.DAT    Bess and Bailey Dog Kennel System
        DOG3.DAT    Process Receptionist Command
Last changed: DEFAULT    19-Dec-2000 18:49:42

----- End of report -----
```

Third refinement: Display information on the receptionists VDU

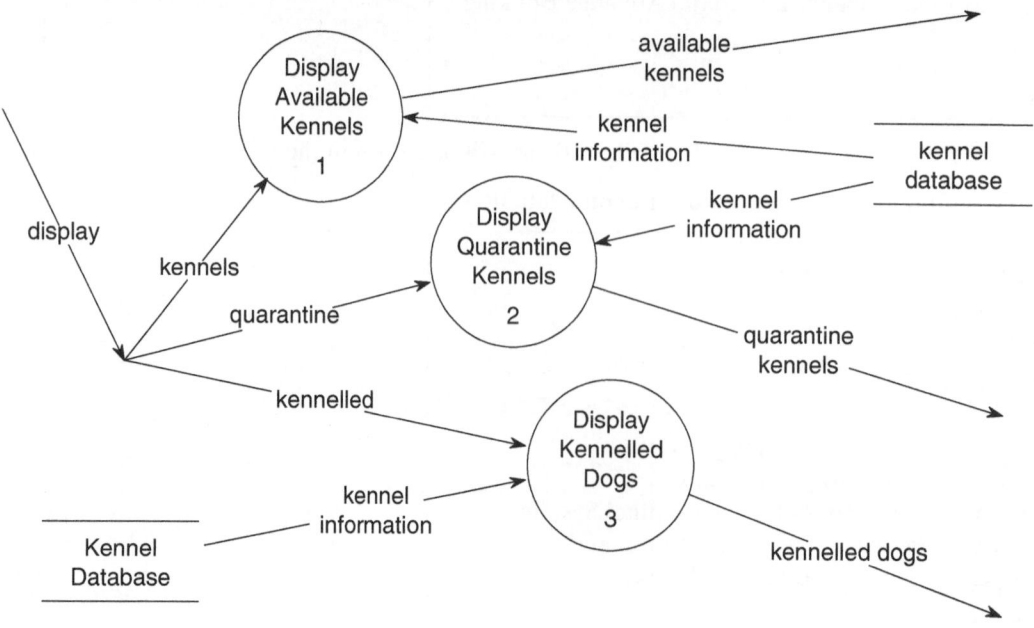

Figure 2.5 *Display Information: third refinement*

Check of the Display Information refinement

> Project: C:\SELECT\SYSTEM\BBDC01\
> Title: Bess and Bailey Dog Kennels
> Date: 19-Dec-2000 Time: 17:0
>
> Checking DOG5.DAT
>
> No Errors detected, No Warnings given.
>
> ----- End of report -----

Group Node requirements

> Project: C:\SELECT\SYSTEM\BBDC01\
> Title: Bess and Bailey Dog Kennels
> Date: 19-Dec-2000 Time: 17:1
>
> Name: **display**
>
> Type: Discrete flow
> Bnf: [kennels | quarantine | kennelled]
> This item is used on the following diagrams:
>   DOG3.DAT   Process Receptionist Command
>   DOG5.DAT   Display Information
> Last changed: DEFAULT   19-Dec-2000 17:04:35
>
> ----- End of report -----

Store Balancing requirements

> Project: C:\SELECT\SYSTEM\BBDC01\
> Title: Bess and Bailey Dog Kennels
> Date: 19-Dec-2000 Time: 17:1
>
> Name: **Kennel Database**
>
> Type: Store
> Bnf: [dog kennel | dog out kennel | kennel information]
> This item is used on the following diagrams:
>   DOG3.DAT   Process Receptionist Command
>   DOG5.DAT   Display Information
> Last changed: DEFAULT   19-Dec-2000 17:04:57
>
> ----- End of report -----

Process Specification for the Display Available Kennels process

> @IN = kennel information
> @IN = kennels
> @OUT = available kennels
>
> @PSPEC 0.2.4.1 **Display Available Kennels**
>
> On receiving the kennels display command do:
>   obtain the kennel information from the Kennel Database
>   and determine the available kennels for display on the VDU screen
>
> @

Process Specification for the Display Quarantine Kennels process

@IN = kennel information
@IN = quarantine
@OUT = quarantine kennels

@PSPEC 0.2.4.2 **Display Quarantine Kennels**

On receiving the quarantine command do:
    obtain the kennel information data from the Kennel Database
    and determine the quarantine kennels that have been reserved

@

Process Specification for the Display Kennelled Dogs process

@IN = kennel information
@IN = kennelled
@OUT = kennelled dogs

@PSPEC 0.2.4.3 **Display Kennelled Dogs**

On receiving the kennelled command do:
    obtain the kennel information from the Kennel Database
    and determine the kennelled dogs (i.e. the dogs in the
    kennels at the present time) for displaying on the
    receptionist's VDU

@

Fourth refinement: Print Information to the Receptionists Printer

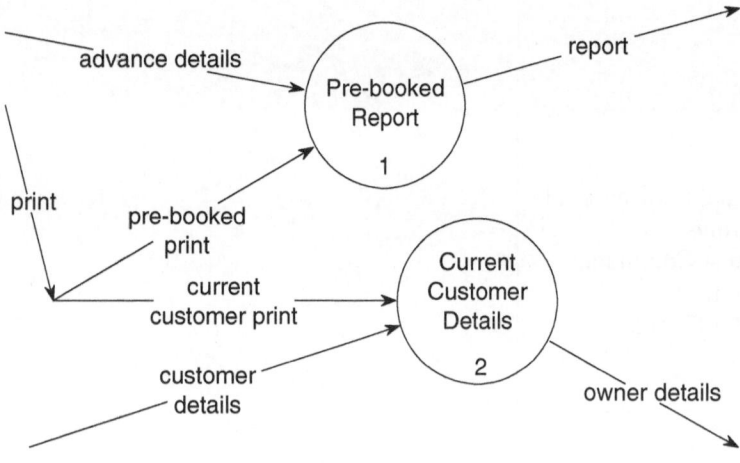

Figure 2.6 *Print Information: fourth refinement*

Check on the Print Information refinement

Project: C:\SELECT\SYSTEM\BBDC01\
Title: Bess and Bailey Dog Kennels
Date: 19-Dec-2000 Time: 17:2

Checking DOG4.DAT

No Errors detected, No Warnings given.

----- End of report -----

Group Node requirements

> Project: C:\SELECT\SYSTEM\BBDC01\
> Title: Bess and Bailey Dog Kennels
> Date: 19-Dec-2000 Time: 17:2
>
> Name: **print**
> Type: Discrete flow
> Bnf: [pre-booked print | current customer print]
> This item is used on the following diagrams:
>   DOG3.DAT   Process Receptionist Command
>   DOG4.DAT   Print Information
> Last changed: DEFAULT   19-Dec-2000 16:50:54
>
> ----- End of report -----

Process Specification for the Current Customer Details process

> @IN = current customer print
> @IN = customer details
> @OUT = owner details
>
> @PSPEC 0.2.5.2 **Current Customer Details**
>
>   On receiving the current customer print command do:
>   obtain the customer details from The Customer Store and
>   produce the owner details of the people who currently
>   have dogs in the kennels (the details to include name, address
>   and telephone number)
>
> @

Process Specification for the Pre-booked Report process

> @IN = advance details
> @IN = pre-booked print
> @OUT = report
>
> @PSPEC 0.2.5.1 **Pre-booked Report**
>
>   On receiving the pre-booked print command do:
>   obtain advance details of people who have pre-booked
>   from the Advance Booking Store
>   and produce a corresponding report to be printed out
>
> @

(iv) Final diagram consistency check

> Project: C:\SELECT\SYSTEM\BBDC01\
> Title: Bess and Bailey Dog Kennels
> Date: 19-Dec-2000 Time: 19:1
>
> Report: Diagram Consistency checking
>
> This report contains a consistency check of all the diagrams in the project.
>
> Checking DOG1.DAT
>   No Errors detected, No Warnings given.
> Checking DOG2.DAT
>   No Errors detected, No Warnings given.

Checking DOG3.DAT
  No Errors detected, No Warnings given.
Checking DOG4.DAT
  No Errors detected, No Warnings given.
Checking DOG5.DAT
  No Errors detected, No Warnings given.
Checking PDOG10.DAT
  No Errors detected, No Warnings given.
Checking PDOG11.DAT
  No Errors detected, No Warnings given.
Checking PDOG4.DAT
  No Errors detected, No Warnings given.
Checking PDOG5.DAT
  No Errors detected, No Warnings given.
Checking PDOG6.DAT
  No Errors detected, No Warnings given.
Checking PDOG7.DAT
  No Errors detected, No Warnings given.
Checking PDOG8.DAT
  No Errors detected, No Warnings given.
Checking PDOG9.DAT
  No Errors detected, No Warnings given.

----- End of report -----

End of the Bess and Bailey Dog Kennels System.

Suggested solution to Exercise 2.2.2 – College MIS System
(i) Context diagram

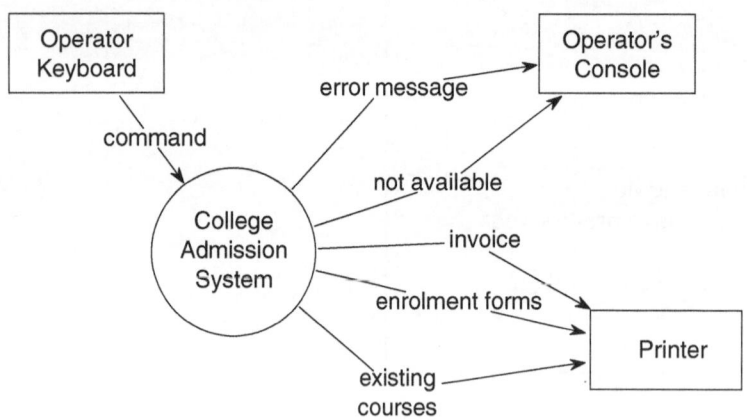

Figure 2.7 *College Admission System: context diagram*

Check of context diagram

Project: C:\SELECT\SYSTEM\COLLEGE\
Title: College Admission System
Date: 20-Dec-2000 Time: 11:0

Checking COLLEGE1.DAT

No Errors detected, No Warnings given.

----- End of report -----

(i) First refinement – College Admission System

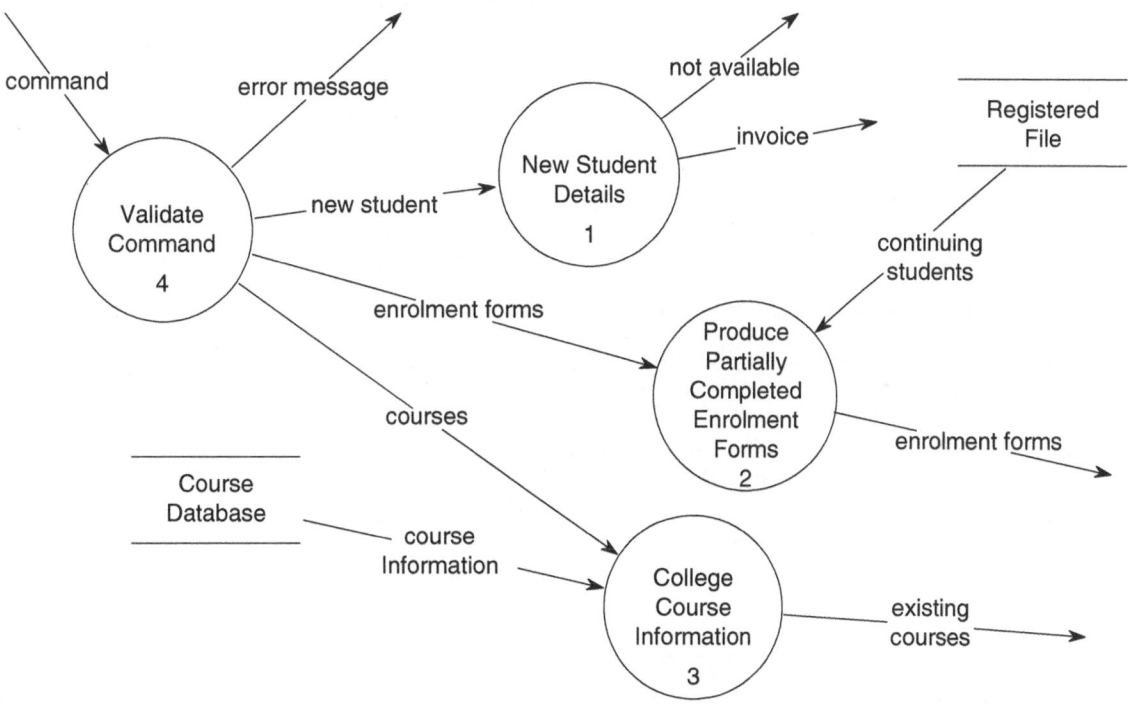

Figure 2.8 *College Admission System: first refinement*

Check for the College Admission System refinement

```
Project: C:\SELECT\SYSTEM\COLLEGE\
Title: College Admission System
Date: 20-Dec-2000 Time: 11:2

Checking COLLEGE2.DAT

No Errors detected, No Warnings given.

----- End of report -----
```

Store Balancing for the Course Database

```
Project: C:\SELECT\SYSTEM\COLLEGE\
Title: College Admission System
Date: 20-Dec-2000 Time: 11:2

Name: Course Database
Type: Store
Bnf: course information
This item is used on the following diagrams:
   COLLEGE2.DAT College Admission System
Last changed: DEFAULT    20-Dec-2000 11:21:04

----- End of report -----
```

Note: The Registered File is also used in the first refinement diagram, but unlike the Course Database this needs to have input information added when new students are registered. Therefore the Store Balancing for this store will be shown after the next refinement.

(ii) Process Specifications for the Validate Command, Produce Partially Completed Enrolment Forms and College Course Information are shown below (these do not require any more refinement)

@IN = continuing students
@IN = forms
@OUT = enrolment forms

@PSPEC 0.2 **Produce Partially Completed Enrolment Forms**

On receiving the forms command from the operator do:
  obtain continuing students information from the
  Registered File and produce partially completed
  enrolment forms for issuing to the continuing students

@

@IN = course information
@IN = courses
@OUT = existing courses

@PSPEC 0.3 **College Course Information**

On receiving the courses command for the operator do:
  obtain the course information for the Course
  Database and produce lists of existing courses
  to be printed off for appropriate students

@

@IN = command
@OUT = courses
@OUT = error message
@OUT = forms
@OUT = new student

@PSPEC 0.4 **Validate Command**

  On receiving the command from the operator do:
    if the command is not valid then
        display an error message
    else
        accept a valid command from
            courses
      or   forms
      or   new student

@

(iii) Refinement for the New Student Details process

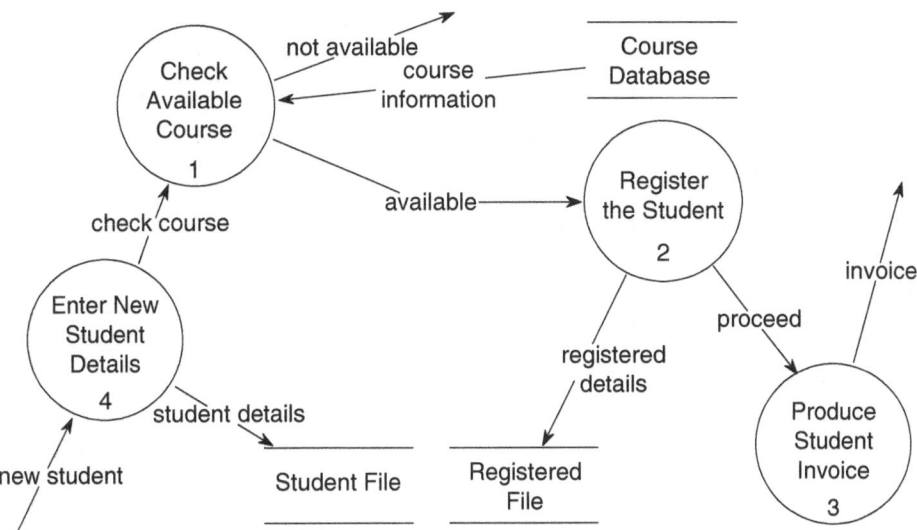

Figure 2.9 *New Student Details*

Refinement check

Project: C:\SELECT\SYSTEM\COLLEGE\
Title: College Admission System
Date: 20-Dec-2000 Time: 12:0

Checking COLLEGE2.DAT

No Errors detected, No Warnings given.

----- End of report -----

Store Balancing for the Registered File

Project: C:\SELECT\SYSTEM\COLLEGE\
Title: College Admission System
Date: 20-Dec-2000 Time: 12:0

Name: **Registered File**
Type: Store
Bnf: [continuing students | student details]
This item is used on the following diagrams:
 COLLEGE2.DAT College Admission System
 COLLEGE3.DAT New Student Details
Last changed: DEFAULT 20-Dec-2000 11:51:33

----- End of report -----

Store Balancing for the Student File

Project: C:\SELECT\SYSTEM\COLLEGE\
Title: College Admission System
Date: 20-Dec-2000 Time: 12:3

Name: **Student File**
Type: Store
Bnf: student details
This item is used on the following diagrams:
 COLLEGE3.DAT New Student Details
Last changed: DEFAULT 20-Dec-2000 12:21:33

----- End of report -----

(iv) Process Specification for the Check Available Course process

```
@IN = check course
@IN = course information
@OUT = available
@OUT = not available

@PSPEC 0.1.1 Check Available Course

        On receiving the check course command do:
            obtain the course information from the Course Database
            if the course exists and there are places then
                    send an available message
            else
                    send a not available message for display on the VDU screen

@
```

Process Specification for the Enter New Student Details process

```
@IN = new student
@OUT = check course
@OUT = student details

@PSPEC 0.1.4 Enter New Student Details

        On receiving the new student command do:
            obtain the new student details and send them to
            the Student File
            then send a check course message to proceed with the registration

@
```

Process Specification for the Register the Student process

```
@IN = available
@OUT = proceed
@OUT = registered details

@PSPEC 0.1.2 Register the Student

        On receiving the available command do:
                obtain the registered details about the student
                and course and send them to the Registered File
                    then
                proceed to produce an invoice

@
```

Process Specification for the Check Available Course process

```
@IN = proceed
@OUT = invoice

@PSPEC 0.1.3 Produce Student Invoice

        On receiving the proceed message do:
                produce an invoice for printing out to the student

@
```

Note: All individual PSPECs should be checked to ensure they are syntactically correct. They must contain the data flows within the body of the explanation.

An example check of the Produce Student Invoice PSPEC is shown below

> Project: C:\SELECT\SYSTEM\COLLEGE\
> Title: College Admission System
> Date: 20-Dec-2000 Time: 13:0
>
> Checking PCAS7.DAT
>
> No Errors detected, No Warnings given.
>
> ----- End of report -----

Which has the same syntax structure as that carried out by the previous diagram checks?

(v) Consistency check of all diagrams

> Project: C:\SELECT\SYSTEM\COLLEGE\
> Title: College Admission System
> Date: 20-Dec-2000 Time: 12:5
>
> Report: Diagram Consistency checking
>
> This report contains a consistency check of all the diagrams in the project.
>
> Checking COLLEGE1.DAT
>   No Errors detected, No Warnings given.
> Checking COLLEGE2.DAT
>   No Errors detected, No Warnings given.
> Checking COLLEGE3.DAT
>   No Errors detected, No Warnings given.
> Checking PCAS1.DAT
>   No Errors detected, No Warnings given.
> Checking PCAS2.DAT
>   No Errors detected, No Warnings given.
> Checking PCAS3.DAT
>   No Errors detected, No Warnings given.
> Checking PCAS4.DAT
>   No Errors detected, No Warnings given.
> Checking PCAS5.DAT
>   No Errors detected, No Warnings given.
> Checking PCAS6.DAT
>   No Errors detected, No Warnings given.
> Checking PCAS7.DAT
>   No Errors detected, No Warnings given.
>
> ----- End of report -----

End of the College Admission System.

SSADM® Analysis and Design

SSADM® is a registered trademark

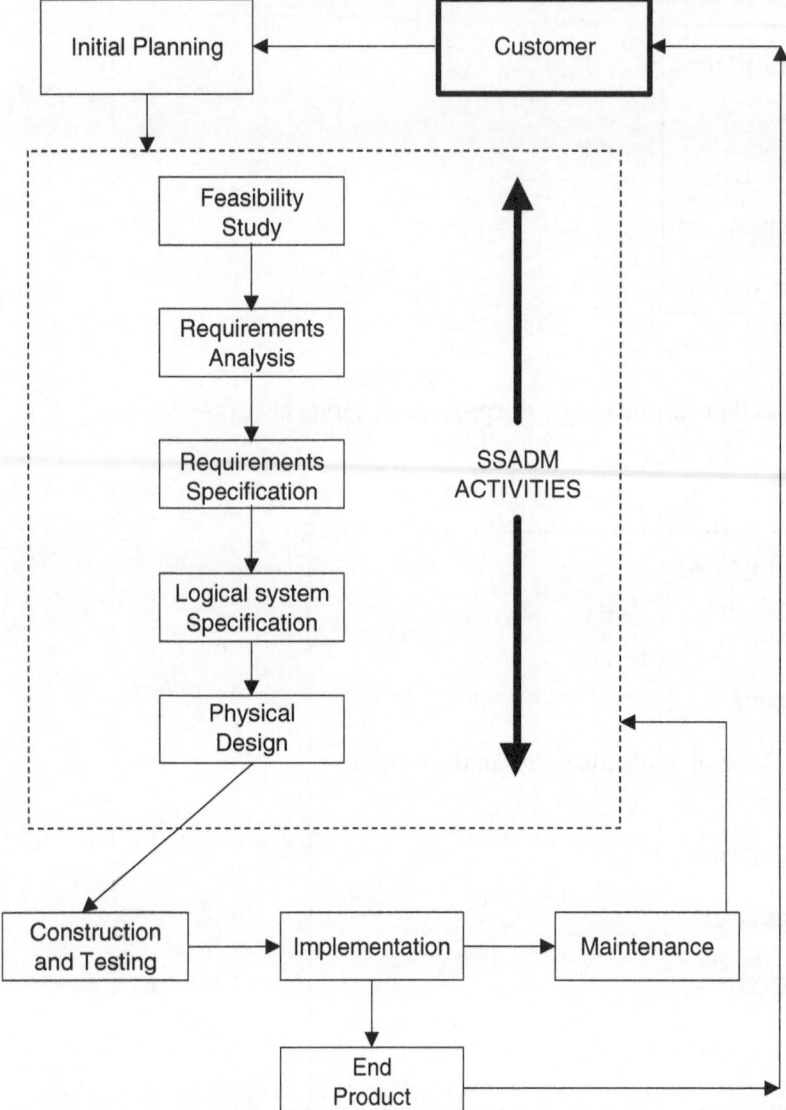

Figure 2.10 *Outline of SSADM (Structured Systems Analysis and Design Method)*

Additional information
Main characteristics
Standard framework for analysis and design:

- A structured framework methodology
- Main tasks include modules, stages, steps and activities.

Standardization of development documentation:

- Standard documentation includes diagrams, forms or reports (SSADM is used extensively within UK government departments and industry, since the release of version 4 the UK government IT agency (CCTA) formalized SSADM into a British Standard
- Provides a common analysis and design strategy, so any person who understands SSADM will be able to review any documentation from any SSADM development process.

Powerful set of development tools and techniques:

- Provides a number of structured techniques to aid the development process, such techniques include:
  - Dataflow Modelling (DFM). This assists the analysis process by producing diagrammatic representations of what the system should do. The diagrams are hierarchical in structure starting with a basic outline (context diagram) and refining through with a greater level of detail
  - Logical Data Modelling (LDM). Here the real life entities and their associated attributes are used to model the files and records within an information system
  - Entity Behaviour Modelling (EBM). An event is an occurrence that changes an entity or associated records in any way. EBM models any unexpected events by introducing concepts of error checking. The view wherein an information system is reflective of processing transactions is handled, for example for a customer's bank account the balance, overdraft values may alter after certain transactions. These events are modelled in an Entity Life History (ELH) diagram that demonstrates the order in which the event may occur.
- Available CASE tools to implement the required structures into a standard format and ensure the syntax meets the requirements of the British Standard (BS 7738).

Once completed the three techniques need to be checked against each other by carrying out rigorous validation checks to ensure the final model is a true representation of the proposed software system.

Table 2.1 shows the permissible links between SSADM symbols.

Table 2.1

	Process	Data store	External entity
Process	yes	yes	yes
Data store	yes	no	no
External entity	yes	no	yes

For example, Figure 2.11 is NOT acceptable.

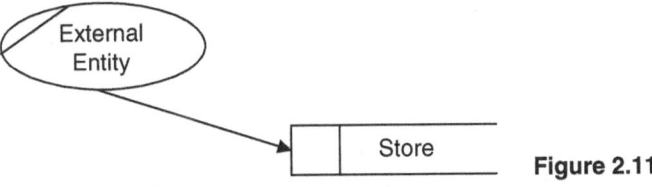

Figure 2.11

Rule: A process must always be present to access a data store.

Note that two external entities can be directly linked. This is shown in Figure 2.12.

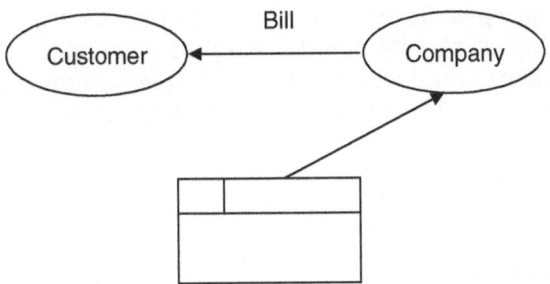

Figure 2.12

Suggested solution to Exercise 2.2.3 – Bess and Bailey Department Store Customer Order System

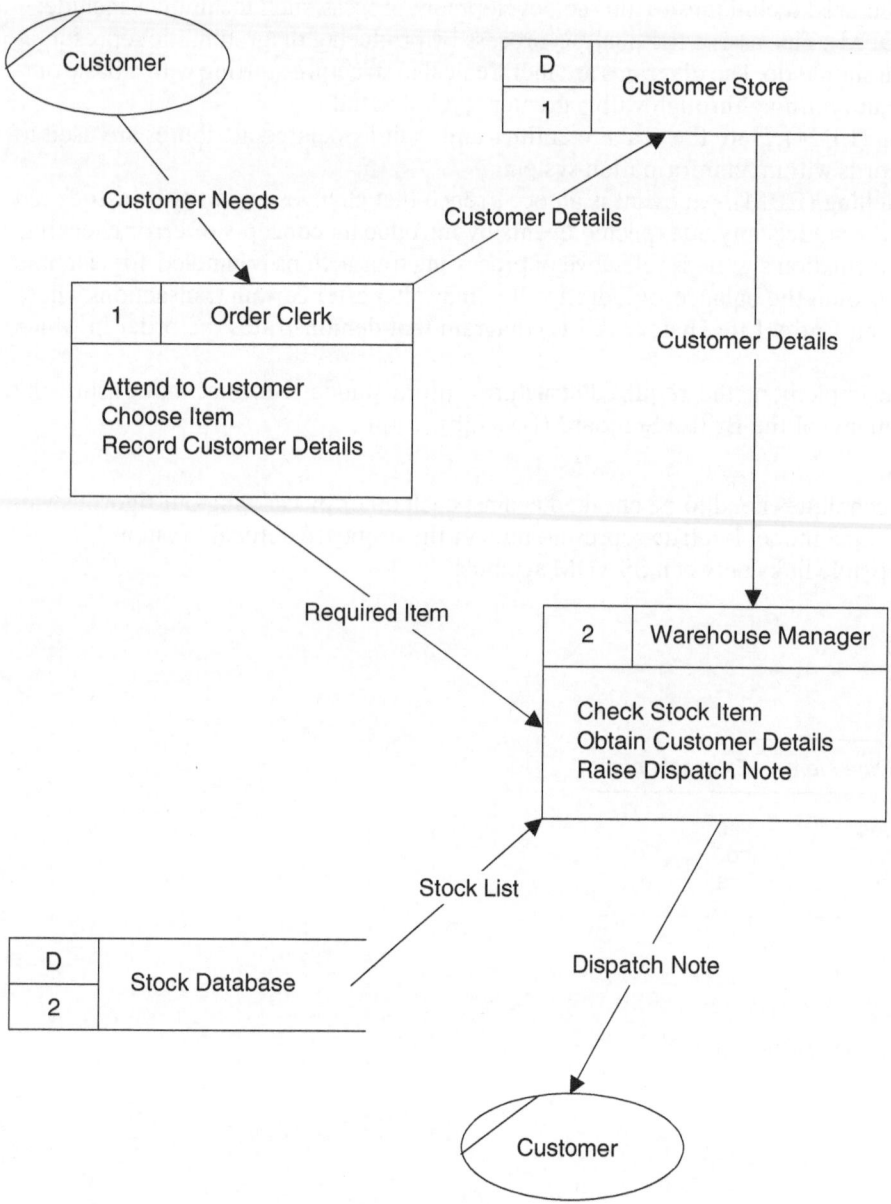

Figure 2.13 *Bess and Bailey Department Store – Customer Order System*

Additional exercise

Bernese Mountain Dog Club Member Event Request System

Outlined below are the system requirements:

The Bernese Mountain Dog Club (BMDC) requires a system to record details from members about the various events that it secludes. Members can make a request for events to be carried out by the club or modifying existing events. Initially all member request are checked by the club secretary, who first checks the member details (i.e. to ensure they are fully paid up members etc.) then records the request for safe keeping. The request is then added as an agenda item for the next committee meeting. The committee members will discuss the request and if adopted it will be stored in the event's file and the member notified. If the request is not adopted then the member is notified to why it has been rejected.

Analyse the above requirements and:

1. Complete an SSADM context diagram
2. Refine the context diagram to fully represent the requirements.

Member
Member
Member Details
and
Event Request
Notification
Bernese
Mountain Dog
System

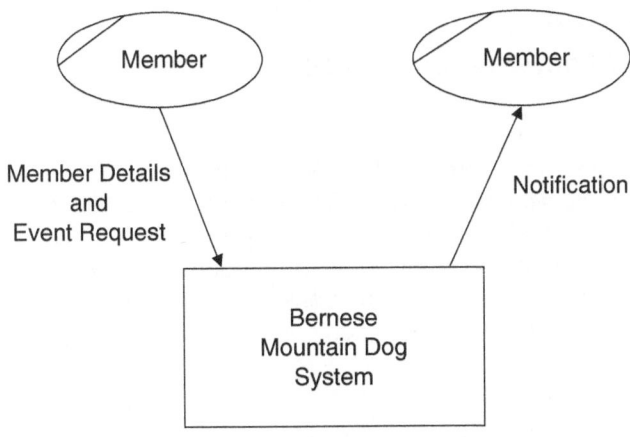

Figure 2.14 *Bernese Mountain Dog Club Member Event Request System: context diagram*

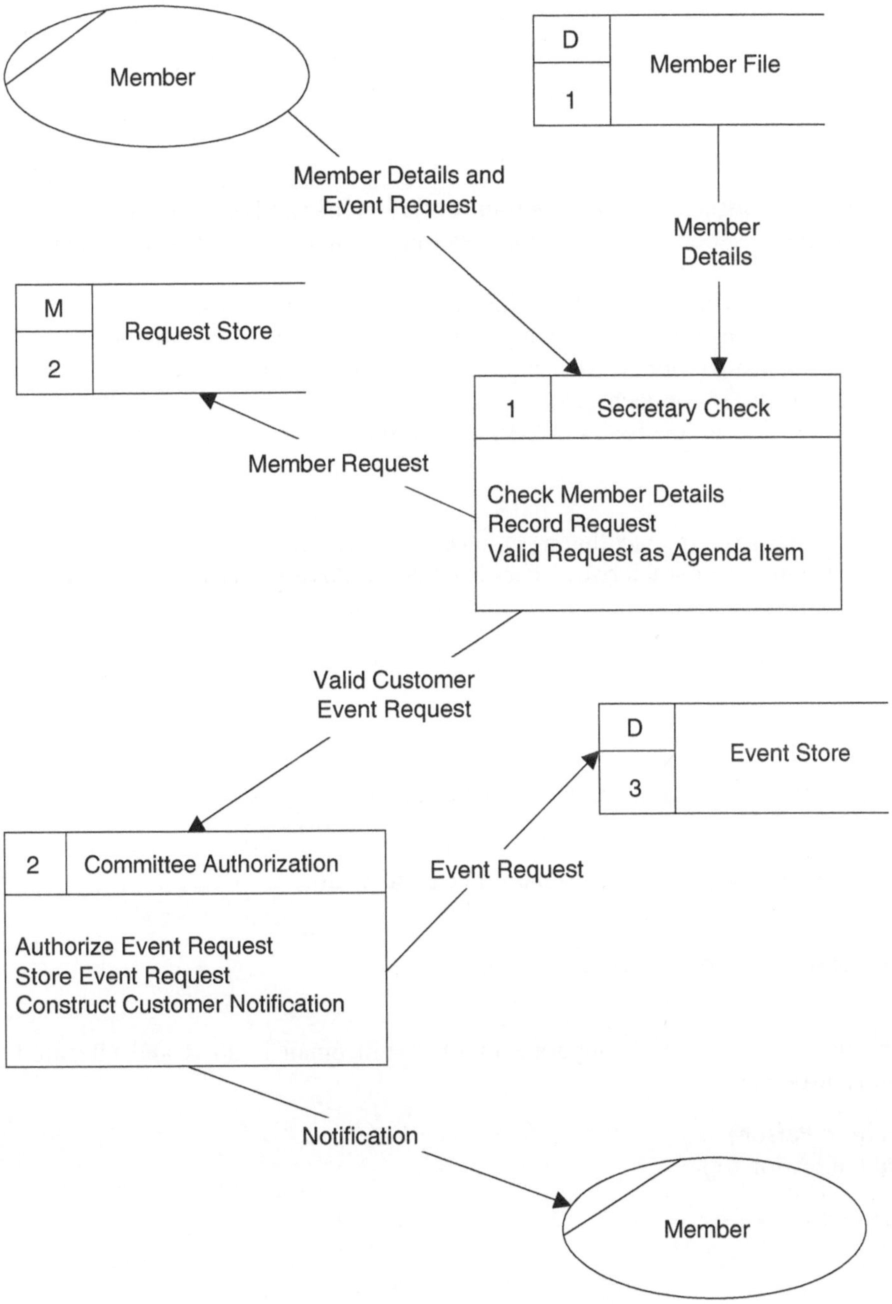

Member
D
1
Member File
Member Details and
Event Request
Member
Details
M
2
Request Store
1
Secretary Check
Member Request
Check Member Details
Record Request
Valid Request as Agenda Item
Valid Customer
Event Request
D
3
Event Store
2
Committee Authorization
Event Request
Authorize Event Request
Store Event Request
Construct Customer Notification
Notification
Member

Figure 2.15 *Refinement of the context diagram*

Object-oriented analysis (OOA) development

Suggested solution to Exercise 2.2.4 – Seminar Organization

(i) Textual Analysis:

An <u>organizer</u> needs to keep and update information on a <u>seminar</u> devoted to canine health. The seminar is made up of several <u>sessions</u>, each run by different <u>speakers</u>. The speakers running the sessions have to produce a <u>booklet</u> for their own slot. Each session has a <u>chairperson</u> and that person is allocated this role only once. Information about the speakers who produce the booklets needs to be recorded; each booklet is written by one speaker only and speakers are only required to produce one booklet. For the seminar presentation purposes, a person can be a chairperson or a speaker.

(ii) The required classes are determined from the textual analysis and shown underlined above. The required classes to be presented are:

- Organizer
- Seminar
- Session
- Speaker
- Booklet
- Chairperson.

(iii) Associations:

organizer and seminar (one to one association as there is only one seminar and one organizer)
seminar and session (one to many association as there is only one seminar, but each seminar can be divided into many sessions)
session and speaker (one to one association as each session is run by different speakers)
session and chairperson (one to one association as each person is allocated this role only once)
speaker and booklet (one to one association as each speaker is responsible for only one session and one booklet only needs to be produced for each slot)
booklet and session (one to one association as all booklets are unique for each session).

(iv) Inheritance:

The only inheritance factor comes from the fact that a person can be a chairperson or a speaker. Both are human and must contain common characteristics. Person could be classified as an abstract parent of chairperson and speaker as no instances of it need to be created (i.e. we do not require a person object).

(v) Associated text

Class: Organizer
'There is only one instance of this class, it is the orchestrating instance'

Class: Seminar
Responsibilities: Record the overall theme of the seminar on canine health

Class: Session
Responsibilities: Record the session topic, venue, date and time

Class: Person
Responsibilities: Record the name, address, telephone number and email address and allocated session of the speaker or chairperson

Class: Speaker (inherits from Person)
Responsibilities: Record the session topic

Class: Chairperson (inherits from Person)
Responsibilities:

Class: Booklet
Responsibilities: Record title of booklet and content information

(vi) Class-association diagram

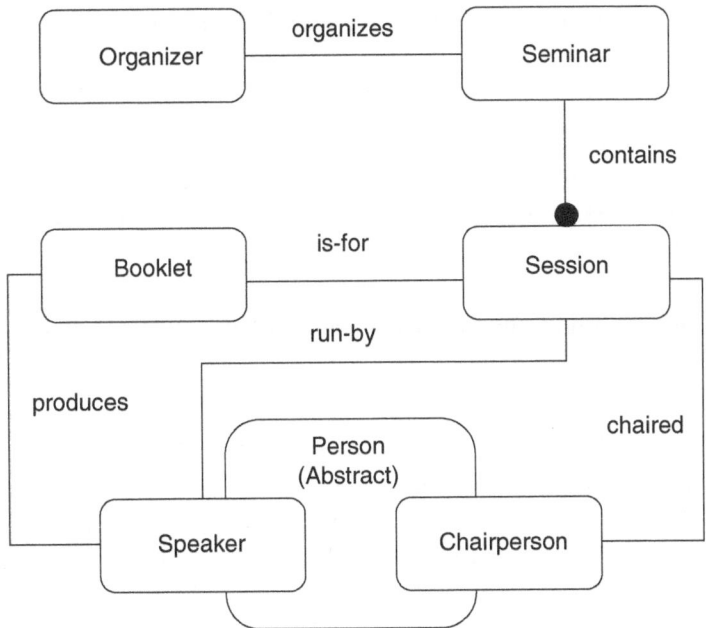

Figure 2.16

Example analysis problem

Ruddles University – Software Engineering Programme

The following is an extract from a negotiated statement of requirements:

> The computing school of the Ruddles University requires a system to keep information about the special Software Engineering Programme (SEP) it is setting up. The programme consists of a number of lectures, each taken by a single lecturer. A lecturer can take more than one lecture slot. Students need to register for each lecture slot and they can attend as many as they want. Students requiring formal certification for the Software Engineering Programme will need to take (and pass) a single examination at the end of the year. One lecturer has the responsibility of writing the examination paper.

Carry out the following analysis of the statement of requirements:

(i)   Perform a textual analysis to ascertain any candidate objects
(ii) Select the required classes to be presented
(iii) Ascertain any associations that exist
(iv) Specify the inheritance factors contained within the text
(v)   Produce associated text to outline the class responsibilities
(vi) Complete a class-association diagram.

Suggested solution

(i)   Textual analysis

> The <u>computing school</u> of the Ruddles University requires a system to keep information about the special <u>Software Engineering Programme (SEP)</u> it is setting up. The programme consists of a number of <u>lectures</u>, each taken by a single <u>lecturer</u>. A lecturer can take more than one lecture slot. <u>Students</u> need to register for each lecture slot and they can attend as many as they want. Students requiring formal certification for the Software Engineering Programme will need to take (and pass) a single examination at the end of the year. One lecturer has the responsibility of writing the <u>examination paper</u>.

(ii) Required classes
- Computing school (Organizer, but not part of the SEP)
- Software Engineering Programme (SEP)
- Lecture
- Lecturer
- Student
- Examination.

(iii) Associations

SEP and Lecture (one to many association as there are many lecturers associated with the SEP)

Lecturer and Lecture (one to many association as each lecture requires only one lecturer, but lecturers can attend more than one lecture)

Lecture and Student (many to many association as many students can attend a single lecture and students can attend many lecturers)

Lecturer and Examination Paper (one to one association as one lecturer is assigned to write a single examination paper)

Student and Examination Paper (one to many association as there is only one examination paper which can be taken by many students).

(iv) Inheritance

As Lecturer and Student share some common characteristics which they could inherit from an abstract class Person

(v) Associated text

Class: SEP

'There is only one instance of this class, it is the orchestrating instance'

Class: Lecture
Responsibilities: Record the title, date, time, location, register and lecturer

Class: Person
Responsibilities: Record the name and email address

Class: Lecturer (inherits from Person)
Responsibilities: Record the room, phone, subject discipline

Class: Student
Responsibilities: Record the course, personal tutor, examination requirements

Class: Examination
Responsibilities: Record the date, time, duration, location.

(vi) Object diagram

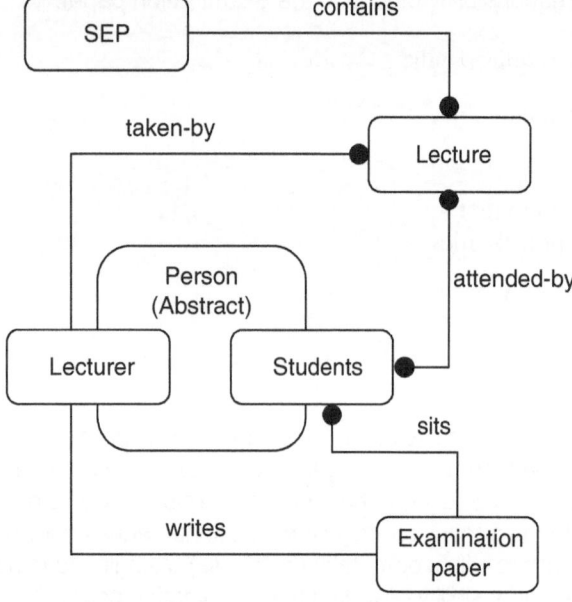

Figure 2.17 *Software Engineering Programme*

Unified Modelling Language (UML)

The UML model provides an industry standard development language which is ideally suited to the design requirements of modern object-oriented languages like Java.

It was adopted by the Object Management Group (OMG) as a standard in 1997.

See:

http://www.omg.org/

Like most other analysis tools UML provides a set of diagrammatic techniques to visualize the relationship of elements between interrelated processes.

UML views
- Use-case view: A view showing the functionality of the system as seen by external actors
- Logical view: A view showing how the functionality is designed inside the system, in terms of the system's static structure and dynamic behaviour
- Component view: A view showing the organization of code components
- Concurrency view: A view that addresses the problems of communication and synchronization that are present in a concurrent system
- Deployment view: A view showing the deployment of the system into the physical architecture with actual computers and associated nodes.

Review of basic diagram structures
Class diagram

Frog
colour: String type: String location: String size: Real

Remember some basic conventions here, for example a lead capital for class names.
Creating two instances of the Frog class gives:

> Frog kermit = new Frog();
> Frog slimey = new Frog();

Creates two Frog objects:

Object diagrams
The convention is to start objects with a lowercase letter.

kermit		**slimey**
colour = "green" type = "Greenback" location = "UK" size = 7.5		colour = "brown" type = "Bullfrog" location = "US" size = 18.75

Suggested solution for Exercise 2.2.5 – Bernese College
(i) Textual analysis:

The Bernese College wants to improve its marketing strategy by setting up an Intranet website primarily to contain a computerized prospectus. The prospectus will contain course information that can be viewed by lecturers, managers, non-teaching staff and students. The college courses are made up of modules that are classified at three different levels; foundation, intermediate and advanced. Students will take three years (full-time) to complete the course starting at the foundation level first. Foundation modules need to be assessed by incorporating level one key skills, Intermediate modules need to be assessed by directed coursework and Advanced modules will incorporate 'A' level examinations.

The prospectus needs to provide a list of lecturers, courses and modules with a more detailed description of the syllabus content of the modules. Information about individual lectures should include their location, main subject disciplines and telephone extension as well as the modules that the lecturer tutors. Each course is allocated a unique identifier and is made up of a number of modules at the required level. The management responsibilities of the prospectus need to add and/or remove lecturers and courses from the system. Each course is allocated a course director, who is also a lecturer, who has the responsibility for running the course and updating the course information contained in the prospectus.

Some of the above candidate identifiers for classes can be ignored as they are not relevant. For example, Students do not need to appear in the prospectus and are not relevant to the analysis model.

The main classes to be considered are:

Website, Prospectus, Lecturer, Course, Module (a super class for the Foundation, Intermediate and Advanced modules)

Note: A Course Director is also a Lecturer (so it does not require a separate class, but its presence will be specified within an association). Courses are allocated a unique identifier (which can locate the course within the prospectus).

(ii) Relationships:
Website *contains* a Prospectus
Lecturer *listed in* a Prospectus
Lecturer (Course Director) *directs* a Course
Lecturer *tutors* a Module
Course *specified in* the Prospectus
Module *contained* in a Course

(iii) Responsibilities:
Prospectus: Update course lists, update lecturer lists, display lecturers, display courses
Lecturer: Record location, subject disciplines, phone extension, current modules
Foundation module: Record level one key skills requirements, foundation syllabus content
Intermediate module: Record directed coursework requirements, intermediate syllabus content
Advanced module: Record 'A' level examination requirements, advanced syllabus content.

(iv) Class diagram

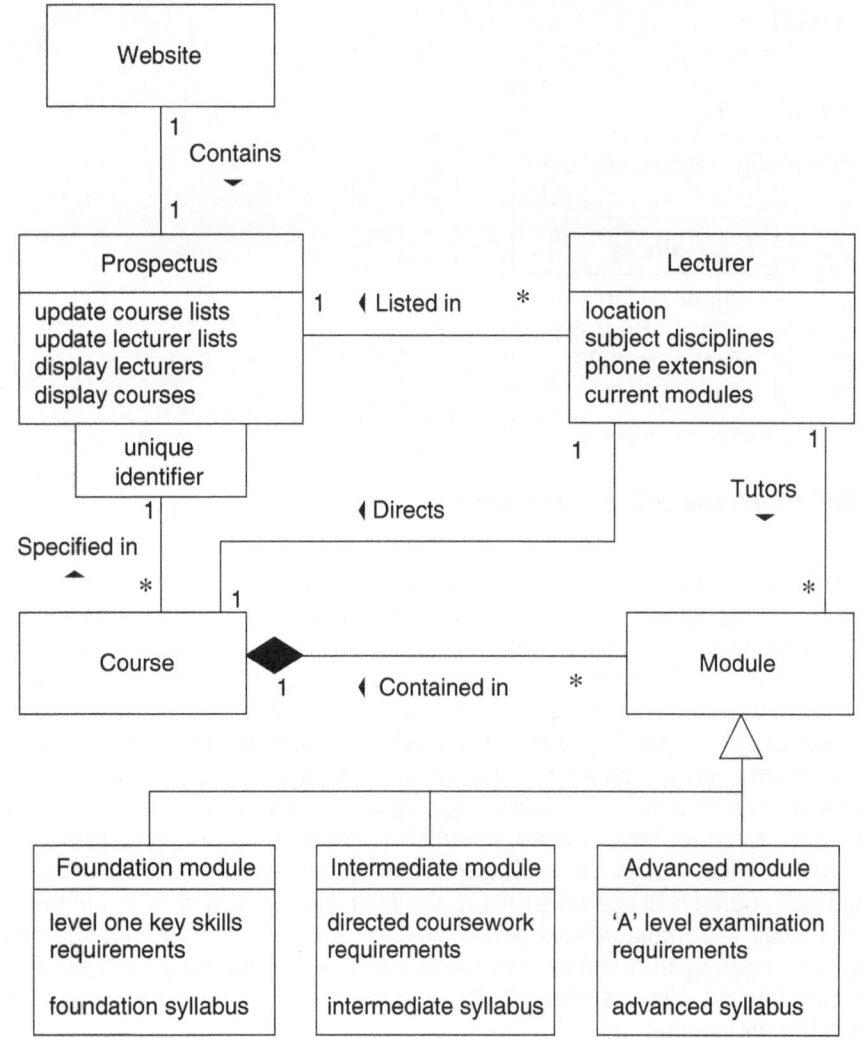

Figure 2.18

First additional UML example

Below is a statement of requirements for the Rio Corner Shop:

> The Rio Corner Shop requires a system to keep information about its newspaper daily delivery service. Each daily delivery involves several rounds, with each round being assigned a unique delivery person. Each round contains customers within given road boundaries. The deliveries consist of different types of newspaper product (local newspapers, national newspapers and news magazines). A customer may order several kinds of newspaper product for delivery and each newspaper product may be ordered by many customers.

Complete the following tasks:

1. Analyse the text to establish the classes to be presented
2. Establish any associations that may exist
3. Specify any inheritance factors contained within the text
4. Produce associated text to outline the class responsibilities
5. Complete a class-association diagram.

Suggested solutions

1. Possible classes (obtained from textual analysis)
   DailyDeliveryService, PaperRound, DeliveryPerson, Customer,
   Newspaper, LocalNewspaper, NationalNewspaper, NewsMagazines
2. Associations
   DailyDeliveryService & PaperRound
   (Each daily delivery consists of many rounds 1:n)
   PaperRound & DeliveryPerson
   (Each round has one delivery person and each delivery person is allocated only one round 1:1)
   PaperRound & Customer
   (Each single paper round has many customers 1:n)
   Customer & Newspaper
   (A customer may order several kinds of newspaper product for delivery and each newspaper product may be ordered by many customers m:n)
3. Inheritance

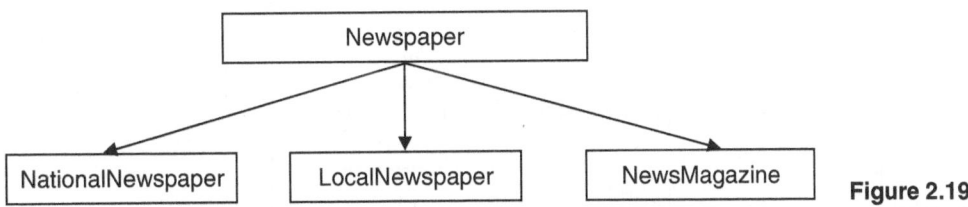

Figure 2.19

   All inherit from Newspaper.
4. Responsibilities
   DailyDeliveryService: An orchestrating instance (only one instance of this)
   PaperRound: Records the delivery name, area and the delivery person
   DeliveryPerson: Records the name, address and telephone number
   Customer: Records the customer name, address and telephone number, delivery requirements
   Newspaper: An abstract class for all paper deliveries. Records the paper name, price
   LocalNewspaper: (Subclass of Newspaper) Records the delivery area
   NationalNewspaper: (Subclass of Newspaper) Records the additional supplements
   NewsMagazines: Records the magazine edition.
5. Class-association diagram to represent the system (see Figure 2.20)

This example is going to look a little deeper into the requirements of the attributes and operations, below is a basic diagram to outline a full class structure:

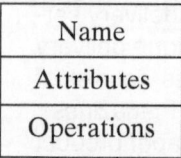

And remember that:

        + expresses public data and
        − expresses private data

The operations will be implemented as methods within a class structure.

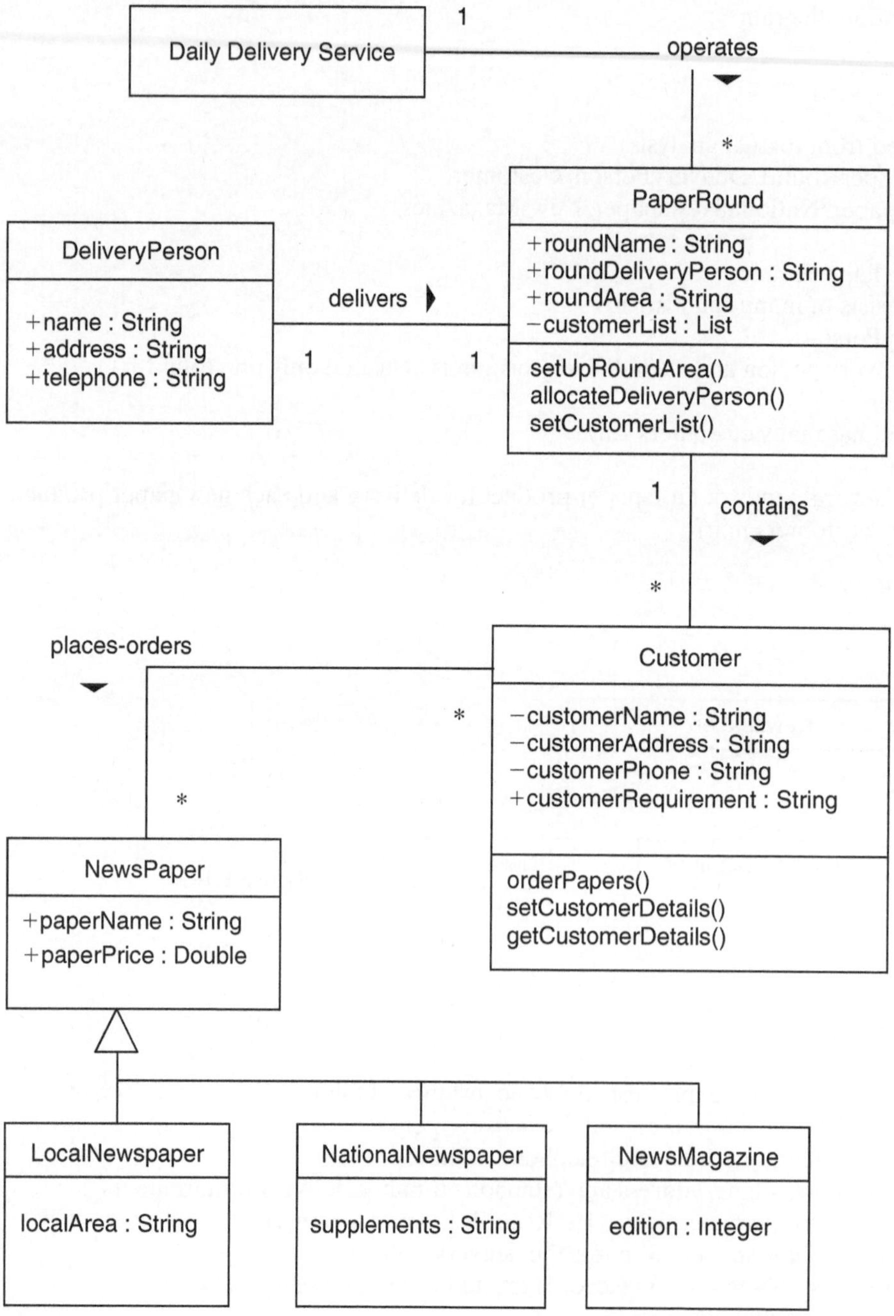

Figure 2.20 *Class-association diagram*

Second additional UML example

Below is part of the statement of requirements for the Remy Conference Company

> An organizer needs to store and update information on a conference to be held by the Carabaz University on program language paradigms. Many individuals are to be involved with the presentations given throughout the conference. The conference consists of several sessions where each individual session involves the presentation of several papers. The papers are unique to each session and will not be issued at other sessions. Each session will have a chairperson, who is selected from the individuals and they will only be asked to chair a single session at the conference. The details of the individuals who are writers of the papers must also be recorded. Note that a paper can have more than one writer and an individual can write more than one paper. For the presentation of the conference an individual is either a chairperson or a writer.

You are to analyse the statement of requirements above and:

- Carry out a textual analysis to find the appropriate classes
- Identify any likely associations between the classes
- Construct a class-association diagram.

Suggested solution

1. Textual analysis: (Note this was carried out through the Visual Paradigm for UML – Community Edition please after this example for further information)

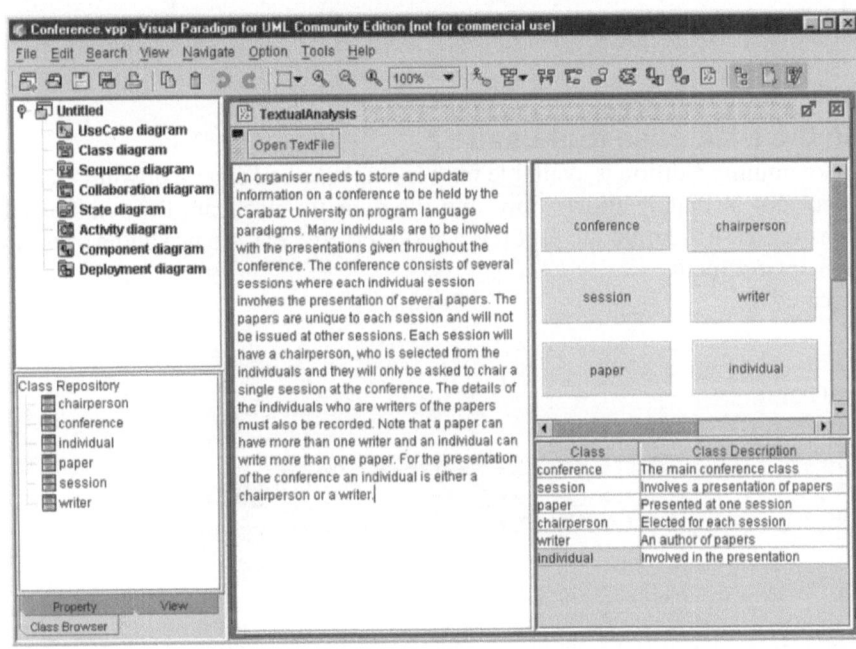

Figure 2.21

The following classes have been identified:

    Conference
    Session
    ChairPerson
    Paper
    Writer
    Individual.

2. Associations

Class	Association	Class
Conference	ConsistsOf	Session
Session	Presents	Paper
Session	HasAn	ChairPerson
Writer	Authors	Paper

3. Class diagram

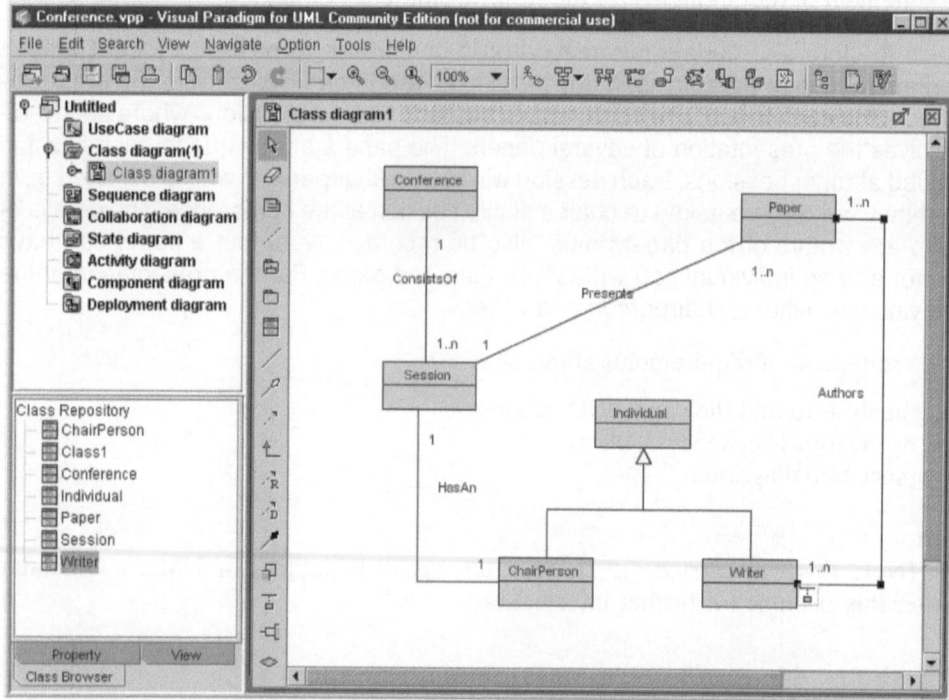

Figure 2.22

Further information on the UML case tool used

Visual Paradigm for UML – Community Edition is available for personal use and can be downloaded free.

The main restriction on the tool is that it allows only one diagram for each application area. For example, one class diagram, one use-case diagram etc. But it is a good tool and easy to use and works along side the JBuilder (Visual Java) environment.

The main website can be located at:

www.visual-paradigm.com

Their home page is outlined below:

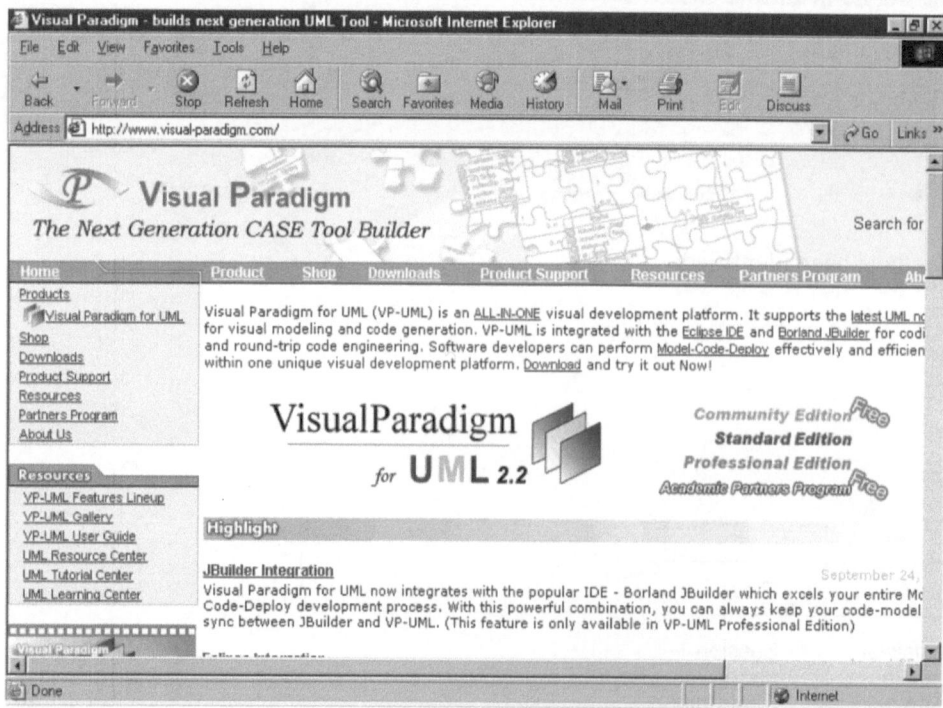

Figure 2.23 *Reproduced by permission of Visual Paradigm International*

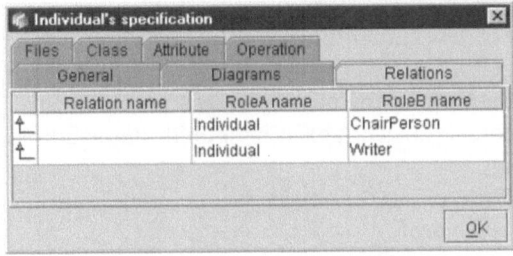

Figure 2.24

All the attribute and operation details can be added to the class along with details of the relations for inherited classes and associations. These are shown under 'Open Specification', click on the 'Relations' tab. It is a complete visual development tool and excellent for developing object concepts using UML notations.

2.3   Systems investigation

Suggested solution for Exercise 2.3.1 – College Management Information System (Admission System)

1. Group size 3 to 4 students
    This is a simulation with one elected person acting as the chair (project manager) and others forming system analysts, programming and technical support etc. roles. One member is elected to take the minutes for the meeting and outline the agenda. It is important that students know about the roles they are taking on. It may be advantageous for the students to research the roles (i.e. what is the main role of a systems analyst? etc.) prior to setting up the committee structure.

    Agenda
    Team members – Introduce each other and specify roles
    Apologies for absence
    Review of the user requirements
    Specify the functional requirements and constraints
    Specify any ambiguous statements, missing components and platitudes
    Any other points of interest.

    Minutes
    These should reflect the discussions held in the meeting by stating who said what and the suggested solutions and/or any resulting conclusions.

2. Functional requirements:
    *   Details of new students entered
    *   Student details stored on a student file
    *   Course check (to see if it is available)
    *   Enrolment – details entered into the registered file
    *   Display suitable message if course not available
    *   Produce invoice for appropriate course fee
    *   Produce partially completed enrolment forms
    *   Produce course information.

    Non-functional requirements (constraints):
    *   Run on the existing college network
    *   Only 300 kb of disk space is available
    *   System is to respond in less than one second.

    Obvious omissions:
    *   How are incorrect inputs by the operator handled?
    *   When/how is the operator going to select required options?
    *   How is data to be entered (see platitudes below)?
    *   Invoices, courses and enrolment forms are to be printed; these require different paper requirements. How is the operator notified of this? Are different printers required?

    Platitudes:
    *   Be easy for the operator to use.

3. Resulting questions:
   - What precise details are to be entered by the operator for new students?
   - What file structure is required?
   - How is the operator told if the required course is available so that the enrolment can proceed?
   - Is a written message enough to tell the operator that a course does not exist or should some audible sound be added alerting the user?
   - What precise details need to be displayed on the invoice?
   - What precise details need to be displayed on the enrolment form?
   - What precise details need to be displayed on the course printout? Does the cost need to be displayed along with the examination board etc.?
   - How are the past students to be linked to the system in order to print out course information to previous enquiries?
   - What type of user interface is required for the system, for example is a Windows environment required via mouse input or would another input medium be required like a touch screen etc.?
   - What are the operational capabilities of the current Apple network system?
   - Is it really necessary for the system to respond in less than one second. It is not a real time system. A wait of two or three seconds would normally seem appropriate.

There may be many more points to add to the list as there is no one fixed solution. The main requirement here is for students to work in groups, grasp the first principles of analysing the problem and understand the importance of providing a sound base, which is clear, concise and unambiguous, for developing potential computer systems. The rule here is that students first *understand the problem before trying to provide a solution*.

Requirements analysis additional exercise 1
The Bess and Bailey Department Store
Goods Ordering System
Outlined is part of the user requirements for the Bess and Bailey Department Store Goods Ordering System:

The department store requires a system to handle its goods ordering system. Order requisitions are set up from within the various departments and sent to the purchasing office for processing. The purchasing office acknowledges the orders from the departments by date stamping them before checking their contents. If there is a problem with the order, for example the description, quantity required, date etc. are erroneous, the order requisition is returned to the issuing department for amending. If the order requisition is OK the supplier name for the required goods to be ordered is added to the requisition from the Supplier Database. If no supplier exists the requisition is stored in a Pending Database and quotations are produced to send to prospective suppliers. Returned quotations from suppliers are added to the Supplier Database and the most suitable one added to the order requisition. Full details about the completed requisition are added to the Purchase Database. Copies of the completed requisition need to be sent to the following:

- Supplier of the goods
- Department requiring the purchase
- Goods In department.

When the order is in the department store from the supplier the Purchase Database is updated to show that the goods have been received. The system is to be installed on the company's existing VAX mini computer system running under a VMS operating system. It needs to be easy to use and incorporates a small number of well-defined principles.

Required steps
1. Within your allocated groups organize and carry out a meeting to analyse the information. The meeting should have a specified agenda and minutes written.
2. Ascertain from the information the following:
   (i) Functional requirements
   (ii) Non-functional requirements
   (iii) Any obvious omissions
   (iv) Platitudes.
3. Construct a suitable set of questions that can provide a basis for an interview with the customer.

Suggested solution

1. This follows the layout for the previous College Admission System
2. (i) Functional requirements
    - Order requisitions raised from within departments
    - Purchasing department acknowledges the requisitions
    - Purchasing department checks the requisitions
    - Incorrect requisitions returned to the issuing department
    - Obtain supplier name from database
    - Produce quotations for new suppliers
    - Maintain supplier database
    - Full details added to the Purchase Database
    - Completed requisitions sent to various departments
    - Record details of orders received.

   (ii) Non-functional requirements
    - Operate on the existing VAX computer system
    - Run the application under the VMS operating system.

   (iii) Obvious omissions
    - When a requisition is received by the purchasing office and date stamped the department raising the order should be notified that it has been received
    - If no existing supplier exists there may be a time delay for ordering the goods. The department issuing the requisition needs to be informed of this as they may need to make alternative arrangements
    - Full details of the completed requisition need to be specified. For example, what precise information is required to be stored for each order?
    - How is the purchase office informed that the goods have arrived so the purchase order database can be updated?

   (iv) Platitudes
    - The system should be 'easy to use'
    - Incorporates a small number of well-defined principles.

3. Points for discussion
    - What details are required on the order requisition from the issuing department?
    - How is the requisition sent from the store department to the purchase office, i.e. is it electronic or manual?
    - Is there a time period for waiting for quotations to be returned from potential suppliers?
    - What is meant by the 'most suitable quotation'?
    - Would a mouse-driven HCI be the best form of program display?
    - What training needs to be given to the users of the system?
    - Are any security features required for the system (passwords, IDs etc.)?

Requirements analysis additional exercise 2

Remy Nearly Always Open Stores

Management System

Outlined is part of the user requirements for the Remy Nearly Always Open Store Management System:

The company owns many stores situated throughout the UK. Each store contains between 4 and 12 departments. The staff of each store are made up of a store manager, managers for each department and sales staff. There is always at least one sales staff person working in a department under the control of the departmental manager. Every week each department undergoes a stock check and the staffing requirements are recorded for the following week's check in advance. It is the responsibility of two or more departmental managers (one must be from the department that the stocktake is being organized for) and one or more sales staff may also participate, but they must be from the department that is being checked.

The system needs to specify the name and telephone extension of the store manager for a given location within the store. It will also provide further details of the store manager to include his/her employment history. For each department the system has to provide a list of the sales staff (their name, full/part-time status and work number) and the corresponding departmental manager. A stocktaking enquiry will produce, for each department, the date of the next stocktake and the work numbers for the staff involved in the check (both managers and sales staff). The system has to be installed on the company's existing micro network which is running under Windows 98. It needs to have a user friendly interface and all staff should find the system easy to use.

Required steps

1. Within your allocated groups organize and carry out a meeting to analyse the information. The meeting should have a specified agenda and minutes written.
2. Ascertain from the information the following:
   - Functional requirements
   - Non-functional requirements
   - Any obvious omissions
   - Platitudes.
3. Construct a suitable set of questions that can provide a basis for an interview with the customer.

Suggested solution

1. This follows the layout for the previous College Admission System
2. (i) Functional requirements
      - For a given location the system should give the details of the associated store manager. This will include his/her name, work number and details of their previous employment
      - Provide a list of all the sales staff working within a specified location. The list has to contain their name, their work number and state if they are full- or part-time. It will also state the departmental manager they are working under
      - Provide details of the next stocking activity for each department. The details should include the date and work numbers of all the staff involved (both departmental managers and sales staff).
   (ii) Non-functional requirements
      - To be installed on the existing company's micro network
      - To run under Windows 98.
   (iii) Obvious omissions
      - How is data to be inputted to update the files?
      - What validation is required when entering data?
      - What security is there for accessing data?
   (iv) Platitudes
      - It should have a user friendly interface
      - Staff should find it easy to use.
3. Points for discussion
   - What is meant by a given location?
   - Will the department hold a stock check on the same day?
   - How does the system ensure that a departmental manager is available for carrying out a stock check?
   - How does the system ensure that the sales staff are from the correct department when carrying out a stock check?
   - How is the information to be displayed?
   - Is there a need for a printer and/or VDU displays?
   - Is there a maximum number of staff that can be employed within the departments?
   - How are transfers across departments to be handled?
   - What is the procedure for handling promotions (or cover for absence) from sales staff to departmental manager?
   - What format does the telephone extension take?
   - How often are the stock checks to be carried out?

2.4 Functional and data modelling

Solution to Exercise 2.4.1 – Fiddle VAT Company currency exchanger

Problem outline

The Fiddle VAT Holiday Company outlined above wants an additional system to help with currency exchange. This system needs to contain a list of countries that are holiday destinations.

The assistant needs to be able to:

- Select a country from a list and press a button to display the currency of the country and its current exchange rate
- Enter the amount of currency that the customer wants to convert and press a button to show the foreign exchange amount
- Press a button to display the currency of the country selected
- Select a country from a list and display the national flag of that country.

Process descriptions

Select a country – An assistant is to select a country from a predefined list for which exchange rates are to be given. The list is already defined using a list box structure so no error checking is required other than outlining the country selected.

Select country for the exchange rate –

Country	Currency	Exchange rate
France	Euro	1.45779
Germany	Euro	1.45779
Greece	Euro	1.45779
Italy	Euro	1.45779
Portugal	Euro	1.45779
Spain	Euro	1.45779
Switzerland	Swiss Franc	2.26110
Turkey	Turkish Lira	2 575 832

Select amount of cash to be exchanged – This will allow the user to select the cash to be exchanged in multiples of £50.

Calculate amount to be exchanged – This process will calculate the amount of foreign currency for the selected country. This is based on the latest exchange rate figures that are to be updated daily within the code structure.

Display – This will display the amount of exchanged cash that has been exchanged along with the currency for the country selected.

Function implementation

The system is to be implemented into a visual programming environment as its rapid application development tools and resulting user interface suit the company requirements. The table below outlines the main functions and the suggested widgets that could implement them.

Functions	Operational widgets
Required exchange countries	List box
Number of pounds (£) to exchange	Combo box
Display amount of exchanged currency	Edit box
Display the currency of the country	Edit box
Calculation of exchange rates	Button
Exit the interface	Button
Display the required flag	Image
User information	Labels

Implementation

The resulting interface is outlined in Figure 2.25.

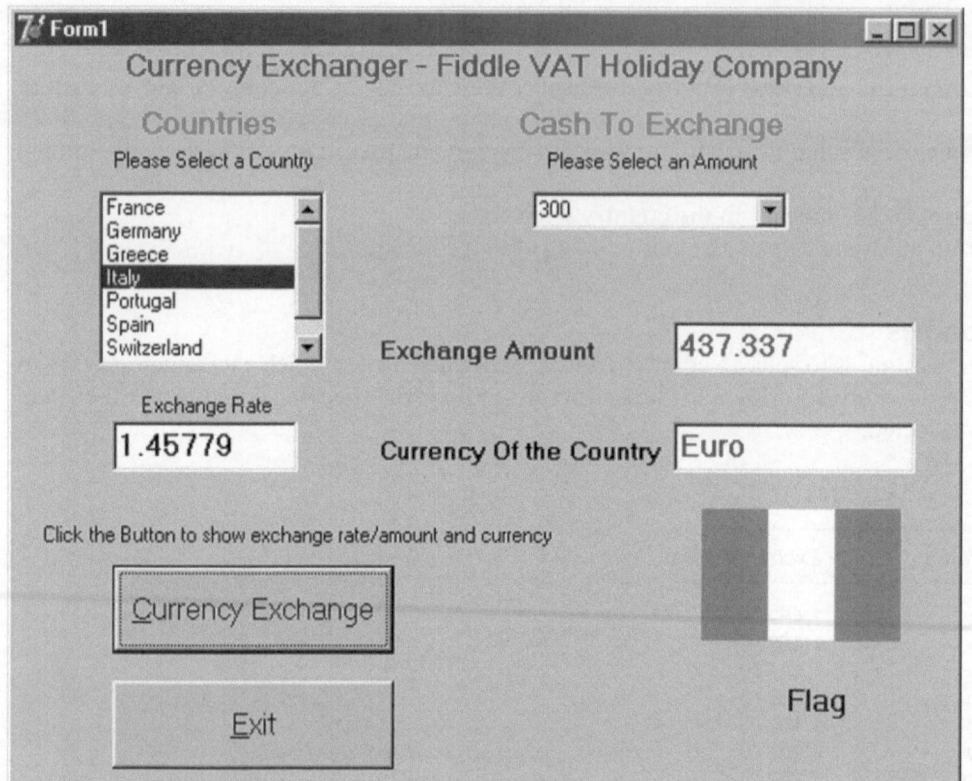

Figure 2.25

The supporting code is:

```
unit Exchange3;

interface

uses
  Windows, Messages, SysUtils, Variants, Classes, Graphics, Controls, Forms,
  Dialogs, StdCtrls, ExtCtrls;

type
  TForm1 = class(TForm)
    ListBox1: TListBox;
    ComboBox1: TComboBox;
    Button1: TButton;
    Button2: TButton;
    Label1: TLabel;
    Label2: TLabel;
    Label3: TLabel;
    Label4: TLabel;
    Label5: TLabel;
    Label6: TLabel;
    Edit1: TEdit;
    Edit2: TEdit;
    Label7: TLabel;
    Label8: TLabel;
    Image1: TImage;
    Label9: TLabel;
    Edit3: TEdit;
    Label10: TLabel;
    procedure Button2Click(Sender: TObject);
    procedure Button1Click(Sender: TObject);
  private
    { Private declarations }
  public
    { Public declarations }
  end;

var
  Form1: TForm1;
```

```
implementation
{$R *.dfm}
procedure TForm1.Button2Click(Sender: TObject);
begin
  Close;
end;
procedure TForm1.Button1Click(Sender: TObject);

var
  exchange : real;
  country : String;

  begin
  { make sure the ComboBox ItemIndex is set to 0 }
  if ( ComboBox1.Text = '0' ) then
    exchange := 0.0
  else
    exchange := strToFloat(ComboBox1.Items[ComboBox1.ItemIndex]);

  { this cannot be set for a listBox hence the check below }
  if ( ListBox1.ItemIndex < 0 ) then
      ListBox1.ItemIndex := 0;
  country := ListBox1.Items[ListBox1.ItemIndex];

  { calculate the currency rate for the country selected }
  country := LowerCase(country);
  if ( country = 'france' ) then
    begin
     Edit1.Text := floatToStr(exchange * 1.45779);
     Edit2.Text := 'Euro';
     Image1.Picture.LoadFromFile('franceflag.bmp');
     Edit3.Text := FloatToStr(1.45779);
    end
else
  if ( country = 'germany' ) then
    begin
     Edit1.Text := floatToStr(exchange * 1.45779);
     Edit2.Text := 'Euro';
     Image1.Picture.LoadFromFile('germanyflag.bmp');
     Edit3.Text := FloatToStr(1.45779);
    end
else
  if ( country = 'greece' ) then
    begin
     Edit1.Text := floatToStr(exchange * 1.45779);
     Edit2.Text := 'Euro';
     Image1.Picture.LoadFromFile('greeceflag.bmp');
     Edit3.Text := FloatToStr(1.45779);
    end
else
  if ( country = 'italy' ) then
    begin
     Edit1.Text := floatToStr(exchange * 1.45779);
     Edit2.Text := 'Euro';
     Image1.Picture.LoadFromFile('italyflag.bmp');
     Edit3.Text := FloatToStr(1.45779);
    end
else
  if ( country = 'portugal' ) then
    begin
     Edit1.Text := floatToStr(exchange * 1.45779);
     Edit2.Text := 'Euro';
     Image1.Picture.LoadFromFile('portugalflag.bmp');
     Edit3.Text := FloatToStr(1.45779);
    end
else
  if ( country = 'spain' ) then
    begin
     Edit1.Text := floatToStr(exchange * 1.45779);
     Edit2.Text := 'Euro';
     Image1.Picture.LoadFromFile('spainflag.bmp');
     Edit3.Text := FloatToStr(1.45779);
    end
```

```
else
  if ( country = 'switzerland' ) then
    begin
      Edit1.Text := floatToStr(exchange * 2.2611);
      Edit2.Text := 'Swiss Franc';
      Image1.Picture.LoadFromFile('switzerlandflag.bmp');
      Edit3.Text := FloatToStr(2.2611);
    end
else
  if ( country = 'turkey' ) then
    begin
      Edit1.Text := floatToStr(exchange * 2575832);
      Edit2.Text := 'Turkisk Lira';
      Image1.Picture.LoadFromFile('turkeyflag.bmp');
      Edit3.Text := FloatToStr(2575832);
    end;
  end;
end.
```

Further test:

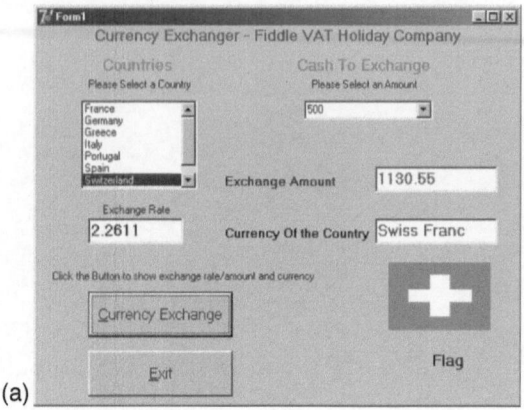 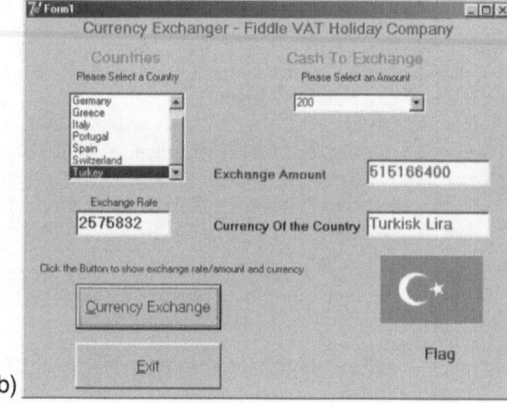

(a)   (b)

Figure 2.26

Data models

A data model will represent the complete set of entities and relationships for a given system. It can be divided into the following stages:

1. A conceptual model of the data representing major objects and relationships
2. A refined model with any redundant components removed
3. Normalized model that is ready for conversion to a physical database.

Normalization has three main aims:

- To find and group together all the properties that are associated with a particular object
- To remove any redundant information. For example, there is no point in storing a person's name, address, telephone number, email address etc. in multiple locations within a database
- To provide unique identification for individual records. For example, you need to make sure when you delete a record from a particular table it is the correct record.

Symbols used:

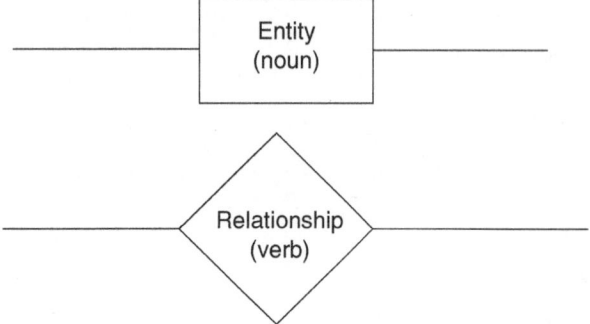

Figure 2.27

Attributes: These represent the 'Data Item Occurrences' that a particular real world entity contains.

Cardinality: Specifies the degree of the relationship (1:1, 1:m, m:1 and m:n).

Quality aspects within database systems

Data Integrity ensures that the overall data quality within a database system is maintained. It refers to the degree to which data:

- Is accurate
- Is available when it is needed
- Is used correctly
- Provides a correct relationship to other data.

Two types of data integrity constraints that need to be evident within a database management system are:

- Inter-record integrity – data types and component values need to be consistent among all entries made in a column of a database table. For example, dates appearing in the same column must always be entered in the same way
- Referential integrity – references between database records must be valid. You need to make sure the referenced data does actually exist.

Approaches to data modelling

1. Top-down design:
   - identifies entities
   - determines attributes of the entities
   - determines the relationships and the associated degrees.
2. Bottom-up design:
   - obtains existing information from within an organization
   - from the attribute data obtained group into required entities
   - determines the nature of the relationships.

The examples shown in the book and the support material use the top-level design as it leads to a data model that is well organized. But this method can lead to problems if important data is left out. The bottom-up method does have the advantage that it helps ensure that no important details are left out. Within industry it is possible that some analysts will use a combination of both methods which can be validated against each other to ensure completeness and sound organizational structure.

Entity-relationship diagrams

These are used as a graphical technique to document the logical design of the database. They can be used to represent the entire logical database scheme or certain user views within the system that contain more detailed information.

Database implementation

The aim of the systems analysis unit is not to teach practical database applications, these are covered by the data analysis and database unit, but to provide the analysis ground work (the conceptual model), develop the design model (logical model) and give a taster of how the model can be implemented.

However, if some students have not yet studied the database unit they will need some basic skills in implementing the analysis/design model.

Microsoft Access lays down the following steps for designing a database:

1. Determine the purpose of your database
2. Determine the tables you need in the database
3. Determine the fields you need in the tables
4. Identify fields with unique values
5. Determine the relationships between tables
6. Refine your design
7. Add data and create other database objects
8. Use Microsoft Access analysis tools.

Step 1 determines the purpose of the database. What are its essential characteristics, how is it going to be used, who is going to use it, what information is required from it and what are the essential characteristics.

Step 2 – the tables can be developed from the entities developed from within the conceptual model.

Step 3 – the fields and their characteristics can be ascertained from the conceptual model attribute diagram where the domains can be used as a base for the required data type.

Step 4 – the primary key defined within the attribute diagram can be used as a unique identifier.

Step 5 – these are contained within the entity-relationship diagram along with their associated degrees.

Step 6 – the logical model will ensure the correct relationship characteristics are maintained by assigning foreign keys. Once the tables, fields, and relationships have been designed, it's time to study the design and detect any flaws that might remain. It is easier to change the database design now, rather than after you have filled the tables with data.

Step 7 – now you can add the data to the tables (and any forms, queries, macros etc.).

Solution to Exercise 2.4.2 – Morgan Software House

(i)   Entities:
Programmer, Terminal, Systems, Office, Manual, Manager

(ii) Relationships:
Manager and Office: 1:1 relationship as each office has only one manager and a manager is only allocated to one office.
Manager and Programmer: 1:n relationship as a single manager is in charge of many programmers and each programmer is responsible to one manager.
Office and Terminal: 1:n relationship as each single office can contain up to 10 terminals.
Programmer and Terminal: 1:1 relationship as each individual programmer is allocated their own terminal and each terminal is used by a single programmer only.
Programmer and System: 1:n (or m:n) relationship as programmers can work on one or more systems. There is no mention in the brief if the systems are worked on by individual programmers or open to many programmers.
Terminal and Manual: 1:1 relationship as each terminal has its own manual and each manual contains the specification of an individual terminal.

Note: Although there must be a relationship between Office and Programmer there is no mention of this in the specification. So it is not included within the list of possible associations.

(iii) Entity-relationship diagram

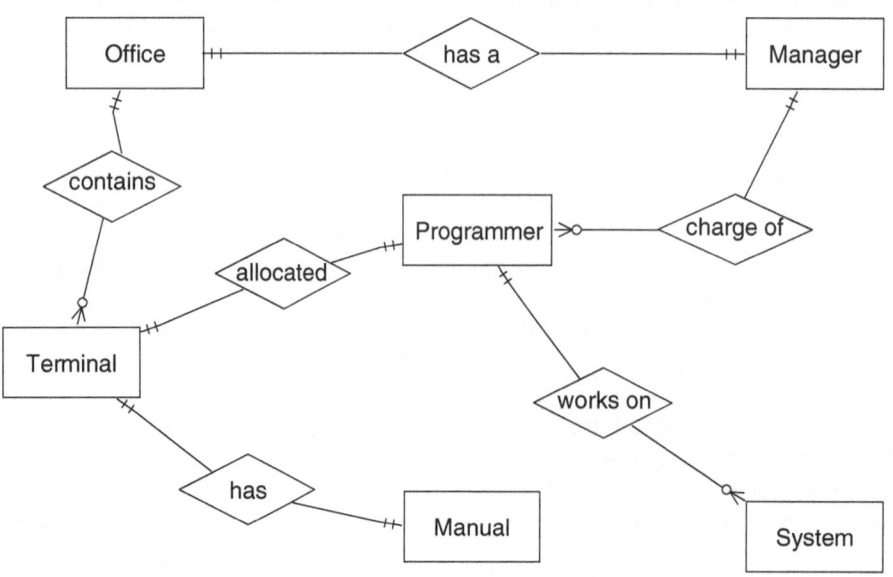

Figure 2.28 *Morgan Software House*

(iv) Attribute diagram – Programmer

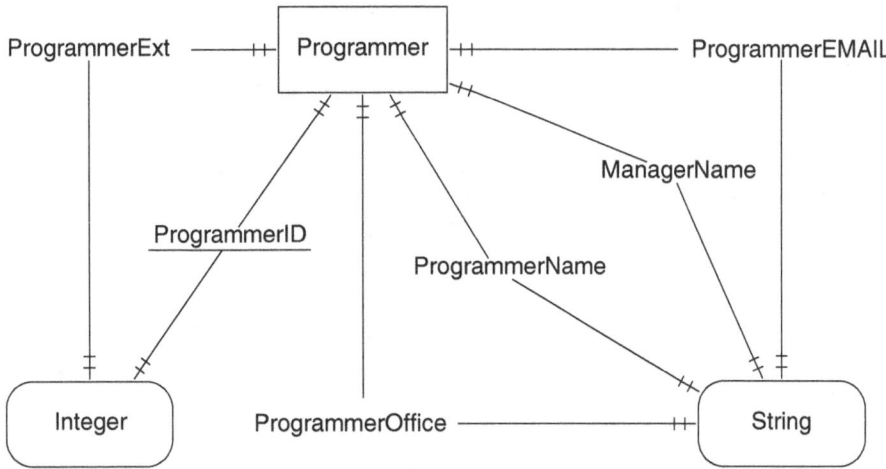

Figure 2.29 *Possible attributes for Programmer*

(v) Attribute diagram – Manager

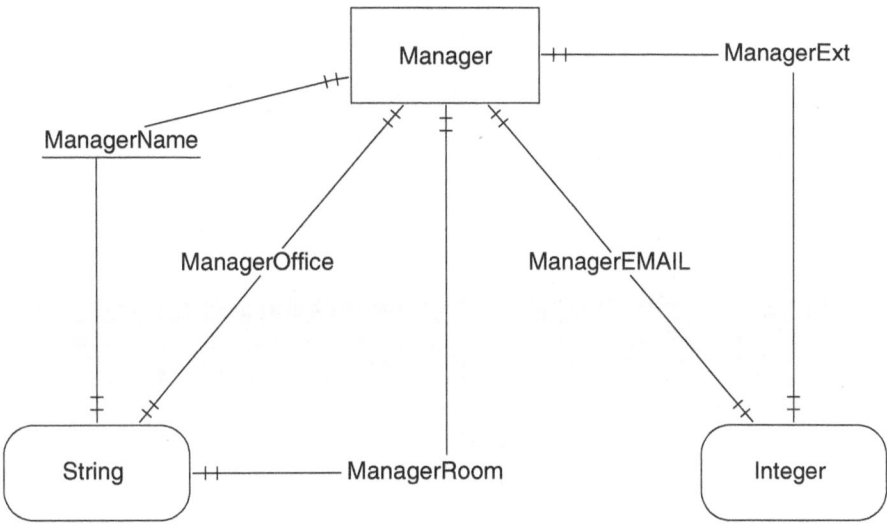

Figure 2.30 *Possible attributes for Manager*

Note: As ManagerName is the primary key of Manager and it has been allocated as a foreign key of Programmer this specifies the 1:n relationship between Manager and Programmer.

Solution to Exercise 2.4.3 – Extract from a College Management System

(i) Entity-relationship diagram

Figure 2.31 *Relationship between Course Director and Student (1:n)*

(ii) Attribute diagram – Course Director

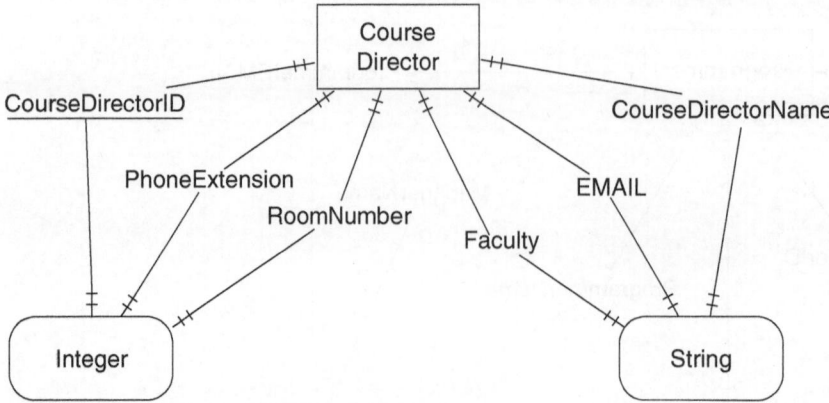

Figure 2.32 *Attribute diagram for Course Director*

(iii) Attribute diagram – Student

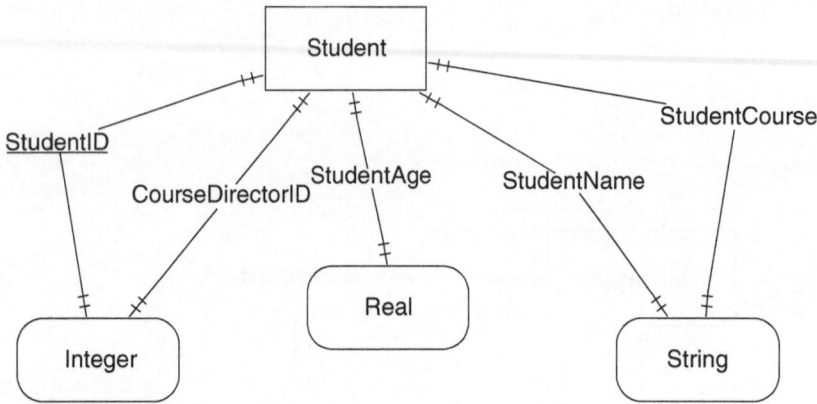

Figure 2.33 *Attribute diagram for Student*

(iv) Database – Student Table

Student ID	StudentAge	StudentName	StudentCourse	CourseDirectorID
13452	16.3	Watson Jill	GNVQ - IT Advanced	3456
15634	16.7	Peters Brian	GCSE - English	6281
16273	19.5	Brown Peter	HNC - Computer Science	5567
23414	24.7	Allison Jane	A Level - French	1963
23452	17.8	Peterson Julie	A Level - History	4653
24152	19.8	Peterson Lee	A Level - History	4653
31093	34.2	Wilson David	HNC - Computer Studies	5567
32000	18.6	Stokes Joan	GNVQ - IT Advanced	3456
34672	22.3	Smith Vivian	HNC - Computer Studies	5567
35465	18.2	Fuller Gillian	HNC - Computer Studies	5567
41243	16.4	Hills Brian	A Level - French	1963
45632	17.9	Khan Ahamdip	GNVQ - IT Advanced	3456
47521	26.8	West Kevin	HNC - Computer Studies	5567
56743	23.5	Coleman Chris	HNC - Computer Studies	5567
65743	16.8	Symonds Katherine	GCSE - English	6281
71342	17.2	Hellings Bess	A Level - History	4653
75643	17.2	Hellings Susie	A Level - History	4653
78654	21.3	Hemmings Gill	HNC - Computer Studies	5567
84391	16.2	Smith Peter	GCSE - English	6281
89654	16.5	Telford Jason	GNVQ - IT Advanced	3456

Record: 1 of 21

Figure 2.34

(v) Database – Course Director Table

CourseDirectorID	CourseDirectorName	RoomNumber	PhoneExtension	EMAIL	Faculty
1963	David Wellingson	231	2674	d.wellingson@fulham.ac.uk	Languages
3456	Julie Heatherington	832	8543	j.heatherington@fulham.ac.uk	IT
4653	Ann Dicks	322	3876	a.dicks@fulham.ac.uk	History
5567	Michael Jaggar	432	4721	m.jaggar@fulham.ac.uk	Computing
6281	John Haynes	576	5322	j.haynes@fulham.ac.uk	English

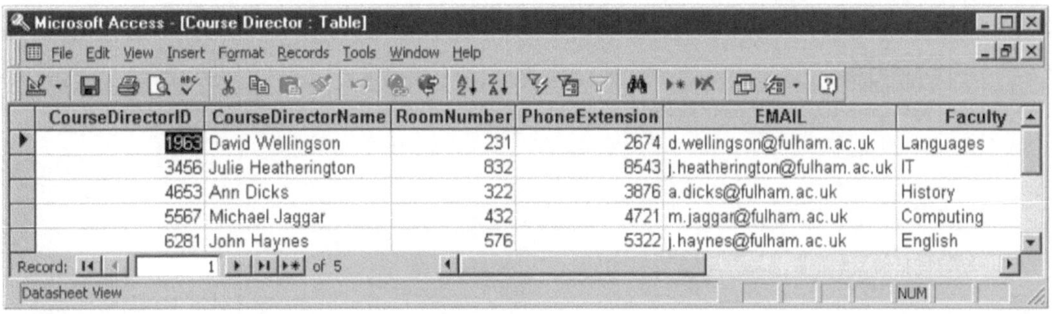

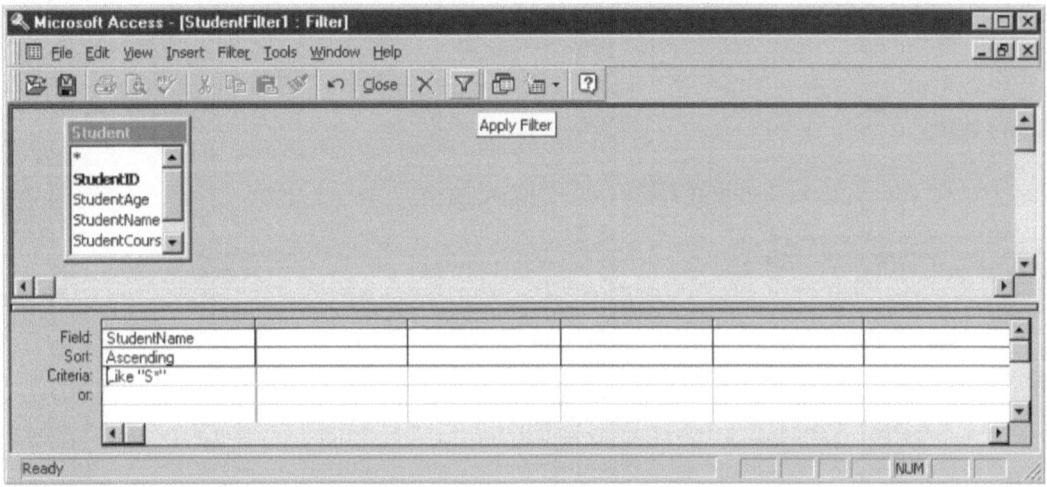

Figure 2.35

(vi) Sorting Course Director Table into Faculty Order

CourseDirectorID	CourseDirectorName	RoomNumber	PhoneExtension	EMAIL	Faculty
5567	Michael Jaggar	432	4721	m.jaggar@fulham.ac.uk	Computing
6281	John Haynes	576	5322	j.haynes@fulham.ac.uk	English
4653	Ann Dicks	322	3876	a.dicks@fulham.ac.uk	History
3456	Julie Heatherington	832	8543	j.heatherington@fulham.ac.uk	IT
1963	David Wellingson	231	2674	d.wellingson@fulham.ac.uk	Languages

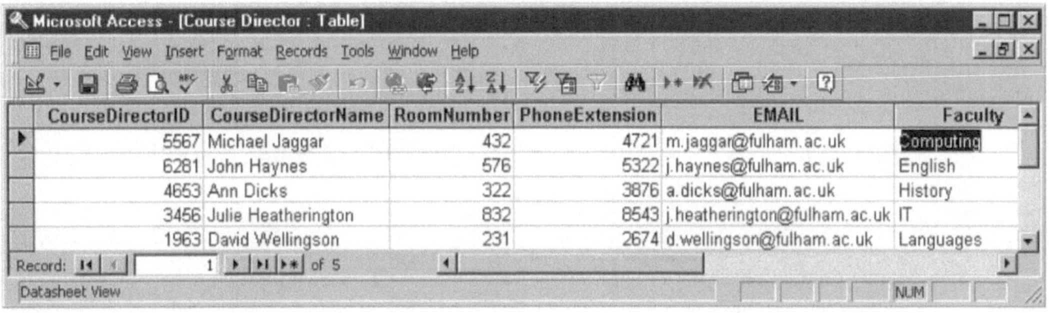

Figure 2.36

(vii) Applying a Filter (selecting: Records, Filter, Advanced Filter/Sort)

Figure 2.37

(viii) Result of the Filter Sort

Student ID	StudentAge	StudentName	StudentCourse	CourseDirectorID
84391	16.2	Smith Peter	GCSE - English	6281
34672	22.3	Smith Vivian	HNC - Computer Studies	5567
32000	18.6	Stokes Joan	GNVQ - IT Advanced	3456
65743	16.8	Symonds Katherine	GCSE - English	6281

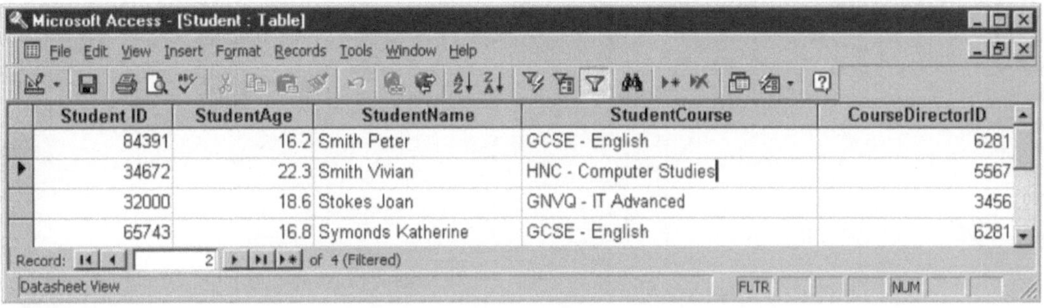

Figure 2.38

(ix)   Form View – selecting a record from the Course Director Table (select the down arrow next to the New Object Button – Just to the left of the help button – then select Autoform and scroll to the required record)

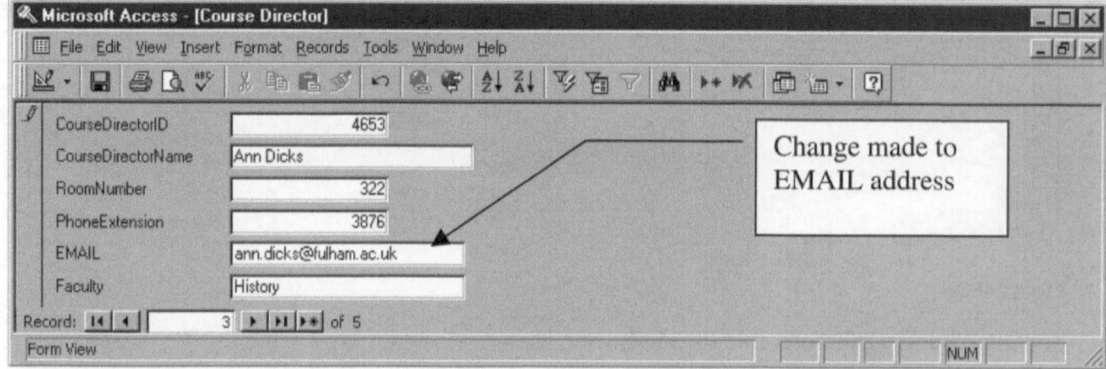

Figure 2.39

(x)   Check on result

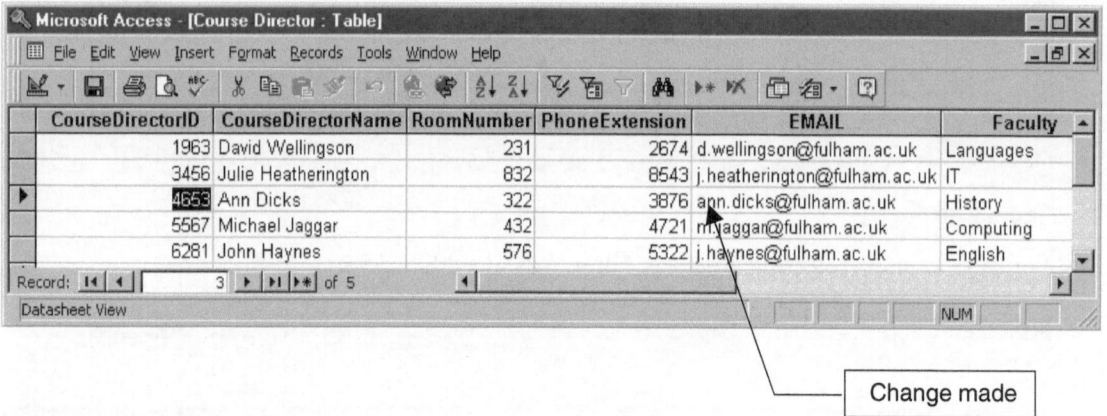

Figure 2.40

(xi)   New record added to the Student Table

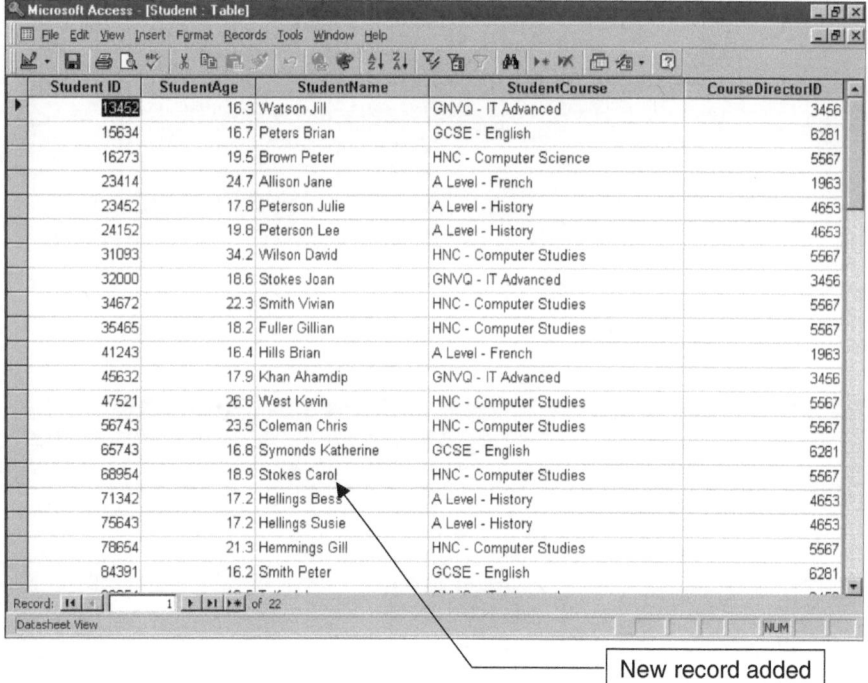

Figure 2.41

(xii) Relationship established

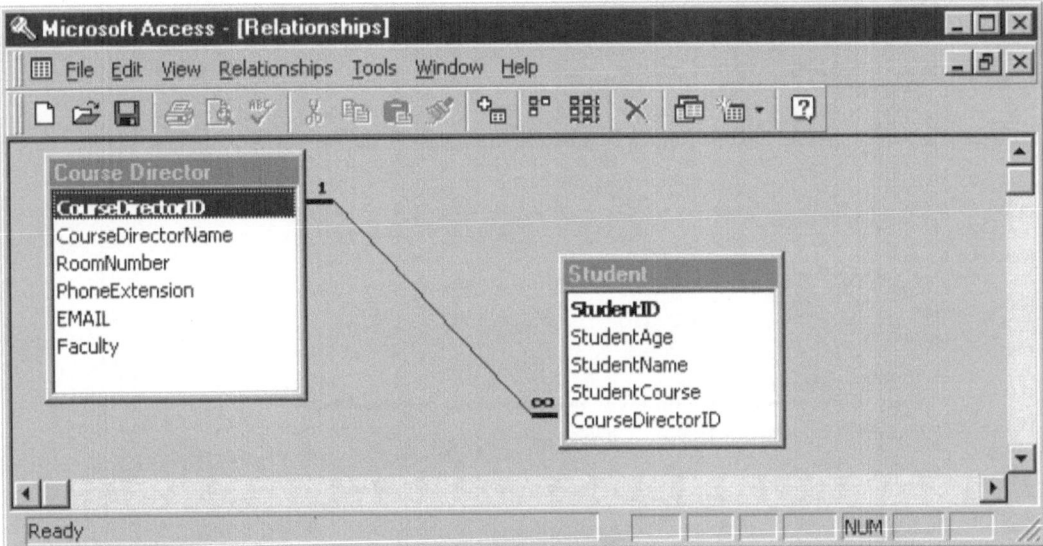

Figure 2.42

Note: As the primary key of the Course Director Table has been assigned as a foreign key of the Student Table the 1:n relationship is established and the relationship is added automatically.

(xiii) Query – Student Names v. Courses

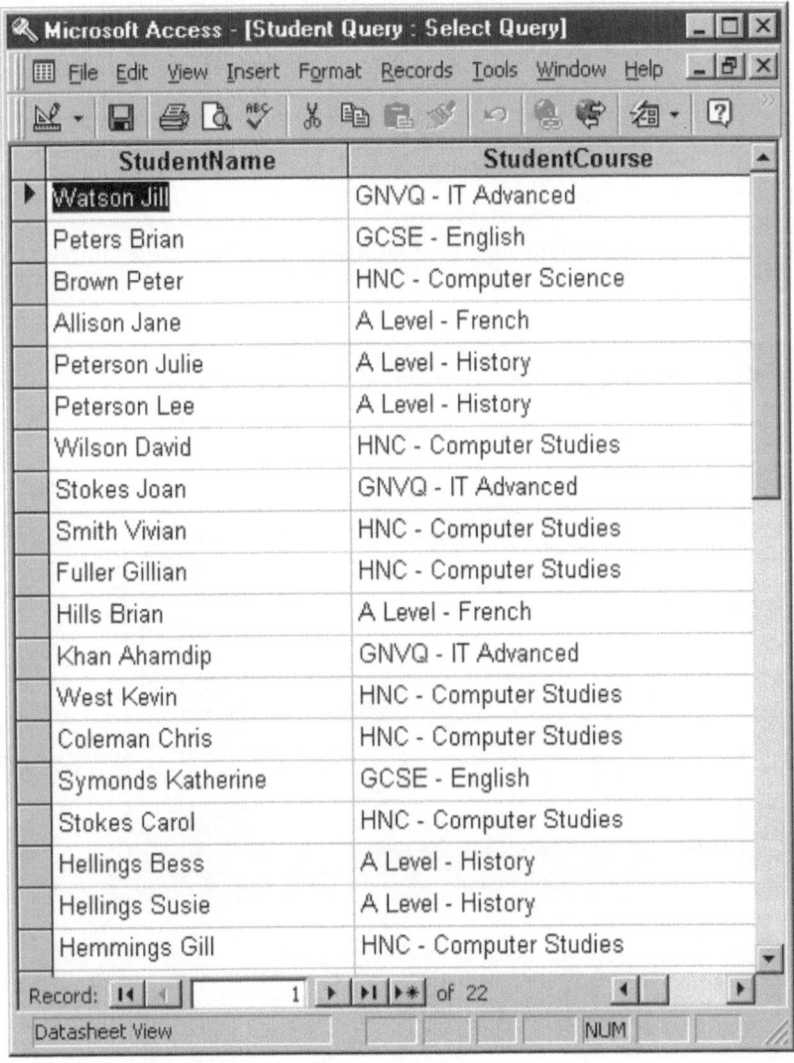

Figure 2.43

(xiv) Report of Course Director's name and EMAIL address

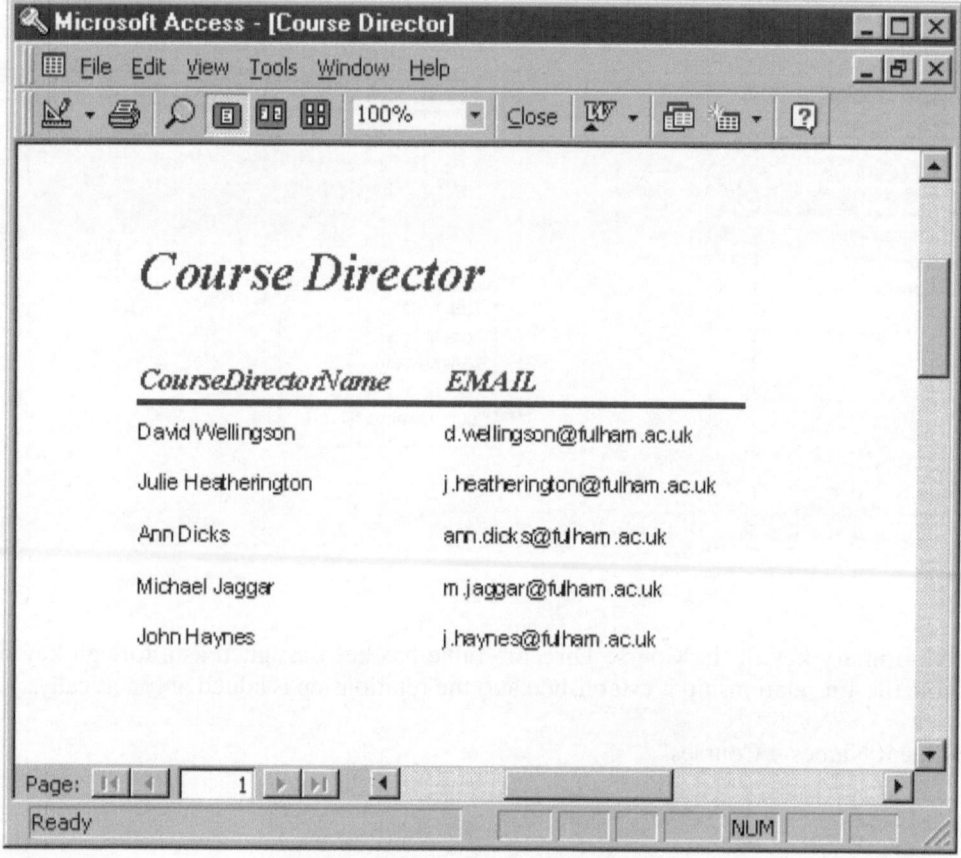

Figure 2.44

Note: Both Queries and Reports are obtained through the Database dialog box. For example, selecting the reports tab gives:

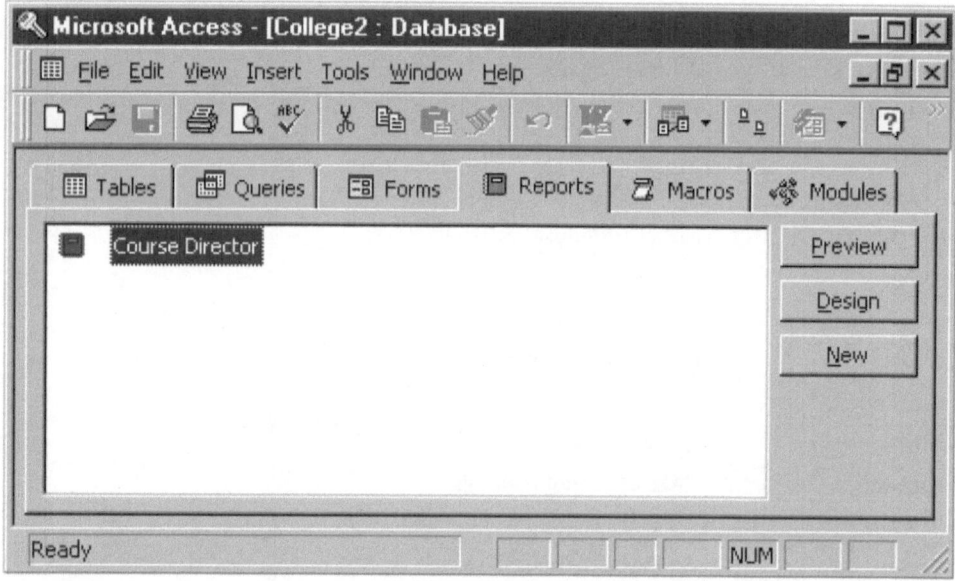

Figure 2.45

You then select New and use the wizard to obtain the required report – i.e. for student names and age:

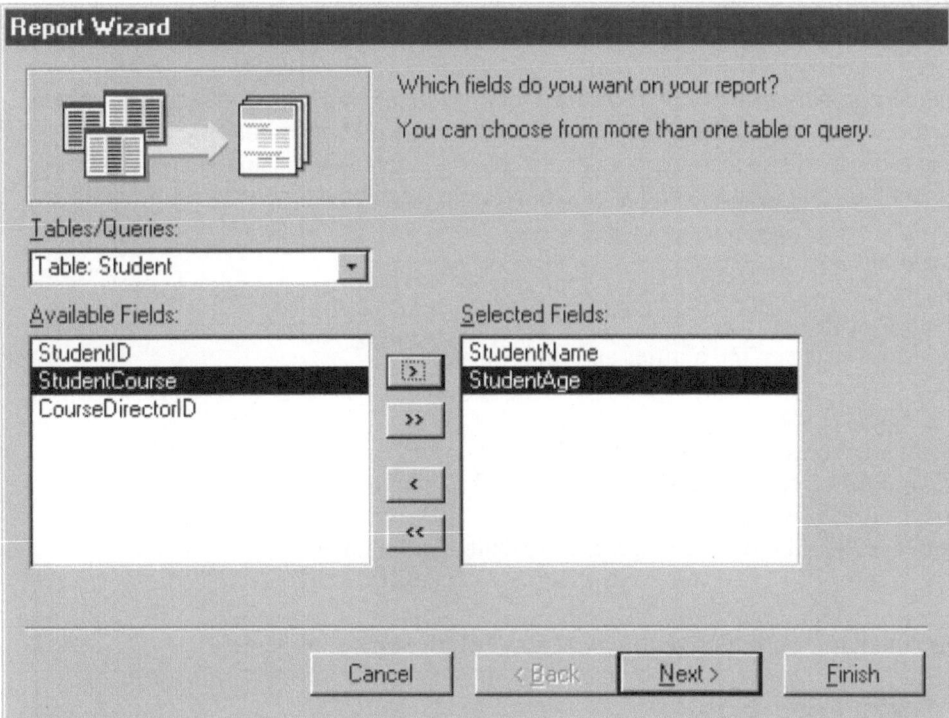

Figure 2.46

Gives the table:

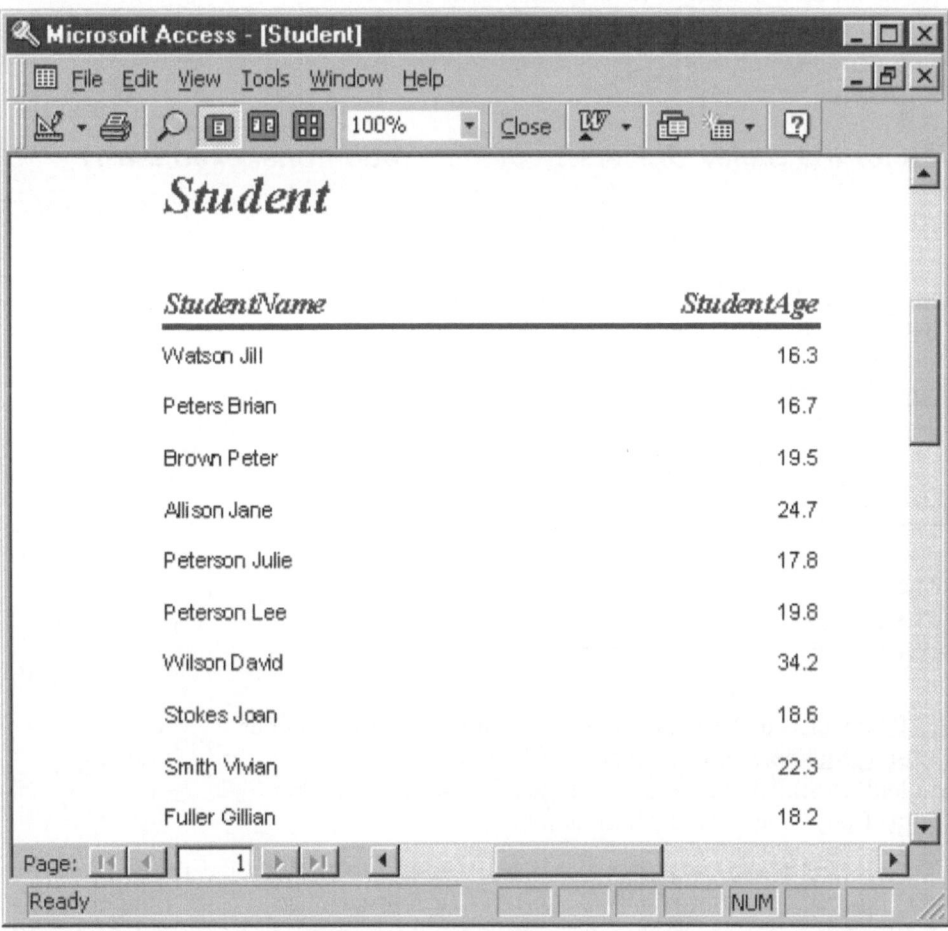

Figure 2.47

Data Modelling – additional exercise 1
Bailey University System

Part of a conceptual model for the Bailey College System is outlined below:

> The university requires all part-time staff to sign a contract for a given tutoring role. Contracts are unique in that each staff member signs one contract and a contract is for one staff member only. The staff are responsible for tutoring more than one student, but each student is tutored by a single member of staff. Students can attend more than one specified course and each course can be attended by many students.

Develop the conceptual model to:
1. Select some suitable entities
2. Outline some attributes for the entities
3. Establish the relationships between the entities and their degrees.

Produce attributes for a logical model:
4. Modify the conceptual attributes to produce a logical model
5. Using a CASE tool application construct an entity-relationship diagram
6. Test the diagram to ensure it is syntactically correct
7. For each of the entities construct attribute diagrams with suitable domains.

Implementation:
8. Construct a database table for the student entity type and include some sample data.

SQL (Structured Query Language)

The final part of this exercise is to use SQL to:
9. Select two columns from the Student Table.
   Using the: Select ....
                From ....
10. Select from two columns within the Student Table given precise field data.
    Using the: Select ....
                 From ....
                 Where ....

Suggested solution for the Bailey University System (additional exercise 1)

1. Suitable entities:
   CONTRACT – STAFF – STUDENT – COURSE

2. Attributes:

Staff	StaffNumber, StaffName, Department, Status
Contract	ContractNumber, ContractType
Student	StudentID, StudentName, StudentAge, Course
Course	CourseCode, CourseTitle, StaffNumber

3. Relationships:

Entity	Relationship	Degree	Entity
CONTRACT	IsFor	1:1	STAFF
STAFF	Tutors	1:n	STUDENT
STUDENT	StudiesA	m:n	COURSE

4. Logical model

STAFF	StaffNumber, StaffName, Department, Status, ContractNumber
CONTRACT	ContractNumber, ContractType, StaffNumber
STUDENT	StudentID, StudentName, StudentAge, Course, StaffNumber
COURSE	CourseCode, CourseTitle, StaffNumber

The identifier of CONTRACT (ContractNumber) has been posted to STAFF and the identifier of STAFF (StaffNumber) has been posted to CONTRACT. As each occurrence of StaffNumber identifies a unique STAFF record and each occurrence of ContractNumber identifies a unique CONTRACT record this represents the 1:1 relationship *IsFor*.

The identifier of STAFF (StaffNumber) has been posted as a foreign key of STUDENT. As each occurrence of StaffNumber identifies a unique STAFF record but may participate in any number of STUDENT records this represents the 1:n relationship *Tutors*.

For fully implementing the design it would be necessary to split the m:n relationship (*StudiesA*) into two 1:n relationships. Here ENROLMENT is introduced for this function.

STUDENT   <u>StudentID</u>, StudentName, StudentAge, Course, StaffNumber
ENROLMENT <u>StudentID, CourseCode</u>
COURSE   <u>CourseCode</u>, CourseTitle, StaffNumber

5. CASE tool representation

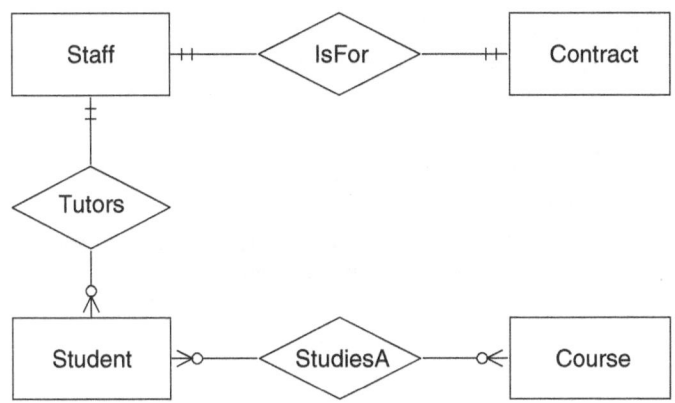

Figure 2.48 *Entity-relationship diagram – Bailey University*

6. Syntax check

Project: C:\SELECT\SYSTEM\BAIL1\
Title: Bailey University System
Date: 10-Apr-2001 Time: 9:2

Checking BAILEY1.DAT

No Errors detected, No Warnings given.

----- End of report -----

7. Attributes

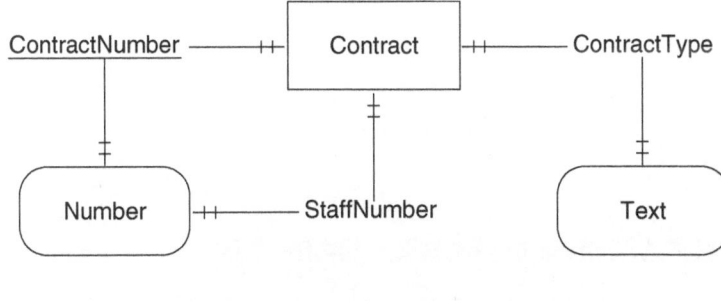

Figure 2.49 *Contract attributes and their domains*

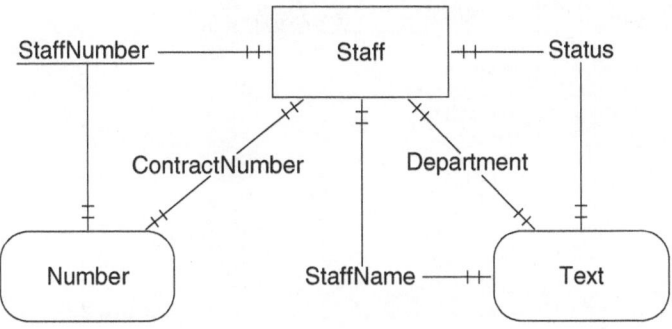

Figure 2.50 *Staff attributes and their domains*

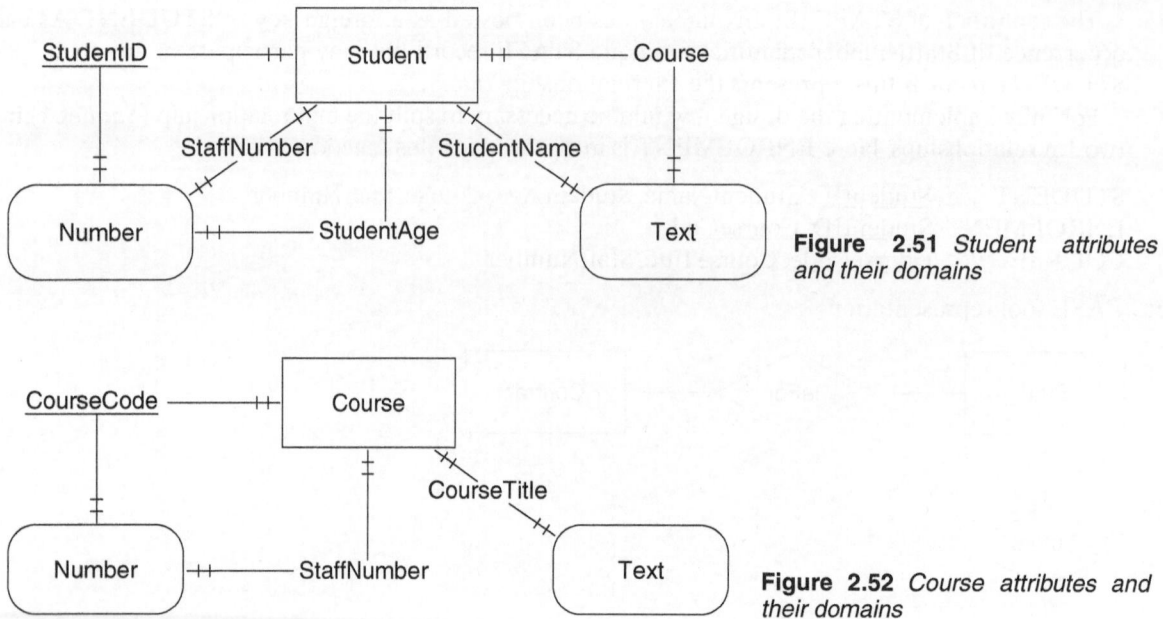

Figure 2.51 *Student attributes and their domains*

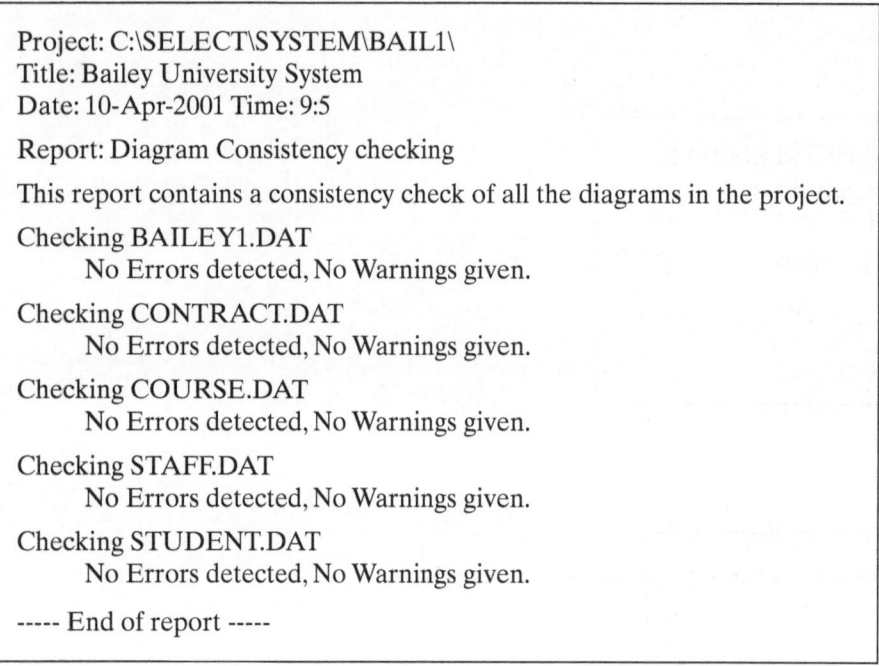

Figure 2.52 *Course attributes and their domains*

CASE tool consistency check

Project: C:\SELECT\SYSTEM\BAIL1\
Title: Bailey University System
Date: 10-Apr-2001 Time: 9:5

Report: Diagram Consistency checking

This report contains a consistency check of all the diagrams in the project.

Checking BAILEY1.DAT
    No Errors detected, No Warnings given.

Checking CONTRACT.DAT
    No Errors detected, No Warnings given.

Checking COURSE.DAT
    No Errors detected, No Warnings given.

Checking STAFF.DAT
    No Errors detected, No Warnings given.

Checking STUDENT.DAT
    No Errors detected, No Warnings given.

----- End of report -----

8. Student Table

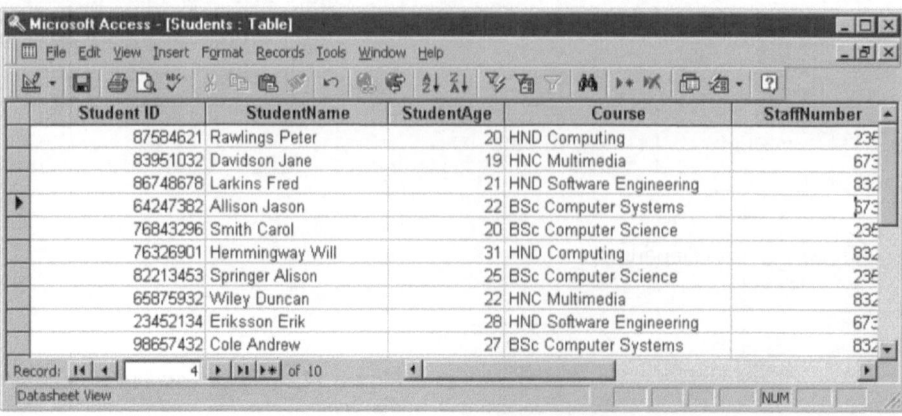

Figure 2.53

9. SQL (Structured Query Language)
   Setting up from the Design View within Queries.

   SELECT StudentName, StaffNumber
   FROM Students;
   produces:

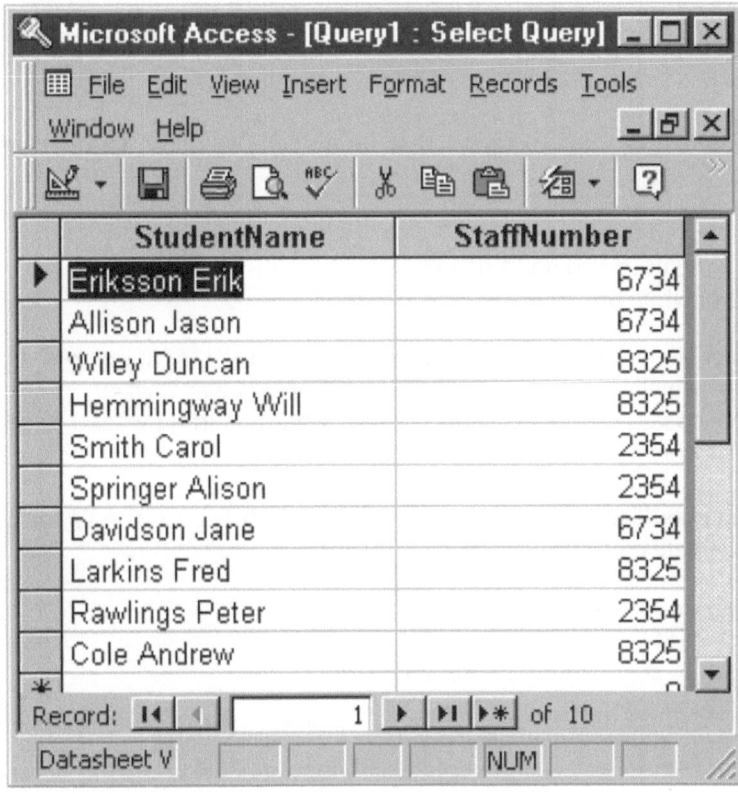

Figure 2.54

10. Using the WHERE statement:

    SELECT StudentName, StaffNumber
    FROM Students
    WHERE StaffNumber = 8325;

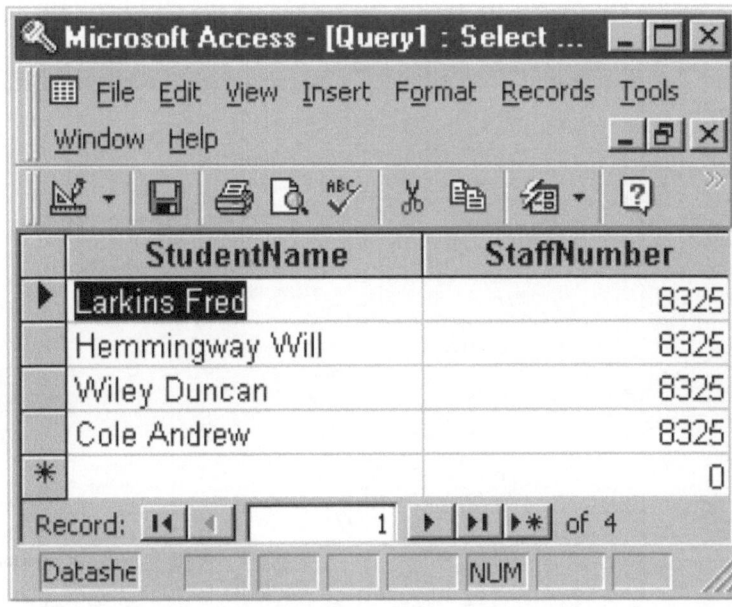

Figure 2.55

Data modelling – additional exercise 2
Morgan Tennis Club
Extract from the negotiated statement of requirements:

The Morgan Tennis Club requires a simple database system for the booking of tennis courts for its members. Information for the following three main areas needs to be considered:

Members
Courts
Booking

Each member who makes a booking is given a booking number for a specified court. The system needs the following:

- Information about the member making a booking
- Each booking needs to record the member and court details
- The court needs to record details of when and how long it is booked for.

Members can make a number of bookings for which a single court is assigned. All members are classified into their respective leagues (which are Premier, Division 1, Division 2 or Division 3).
   The club has 10 courts of various types. They can be grass, sand or hard courts. Two courts are to be allocated to the Premier players only.

Analyse the extract above and complete the following components. Use appropriate case tools for constructing the entity-relationship and attribute diagrams.

(i)   Create an entity-relationship diagram to represent this system
(ii)   Consider some attributes for each entity
(iii)   For each entity create database tables containing some sample data
(iv)   For the member table ensure the names are in ascending order
(v)   Apply a filter to list all the members whose surname starts with a 'C'
(vi)   Display a form view of a particular record for the member table. Change the field entry for one of the attributes
(vii)   Add a new record to the member table
(viii) Show the relationship that exists between the entities
(ix)   Set up a query to produce a subset of one of the tables, for example Members Names v. League Status
(x)   Produce a report for each table
(xi)   Set up a macro to display the Member report, maximized and with a beep.

Suggested solution for the Morgan Tennis Club (additional exercise 2)
(i) Entity-relationship model
   From the question:
   A Member can make several Bookings and a number of Bookings can be made by a single Member (thus establishing a 1:n relationship)
   Many Bookings are made for a specified Court and each single Court can be booked several times (thus establishing a 1:n relationship)

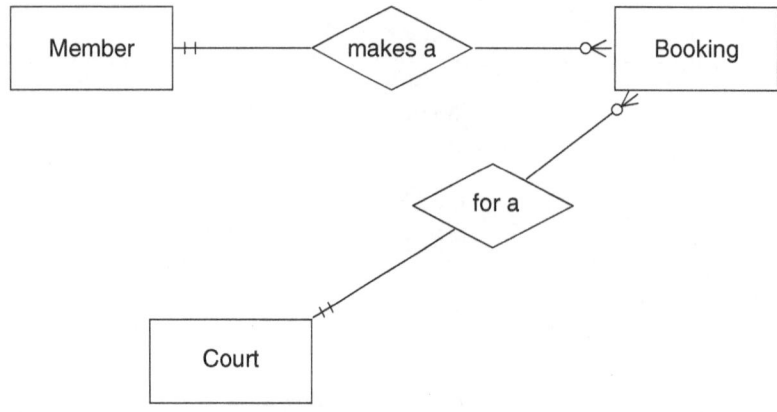

Figure 2.56 *Entry-relationship diagram: Morgan Tennis Club*

(ii) Attributes

Member: (<u>MemberNumber</u>, Name, Address, Telephone, LeagueStatus)
Booking: (<u>BookingNumber</u>, Date, Time, Single, PartnerName, MemberNumber, CourtNumber)
Court: (CourtNumber, CourtName, CourtType, Allocation, InService)

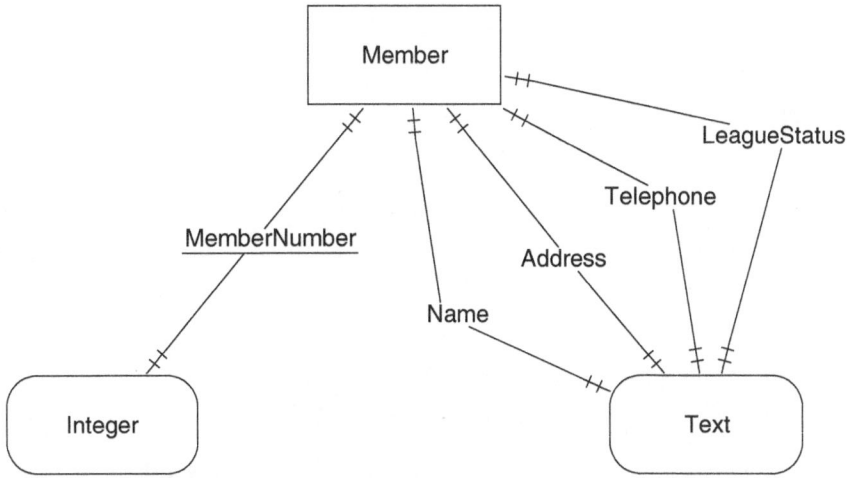

Figure 2.57 *Attribute diagram: Member*

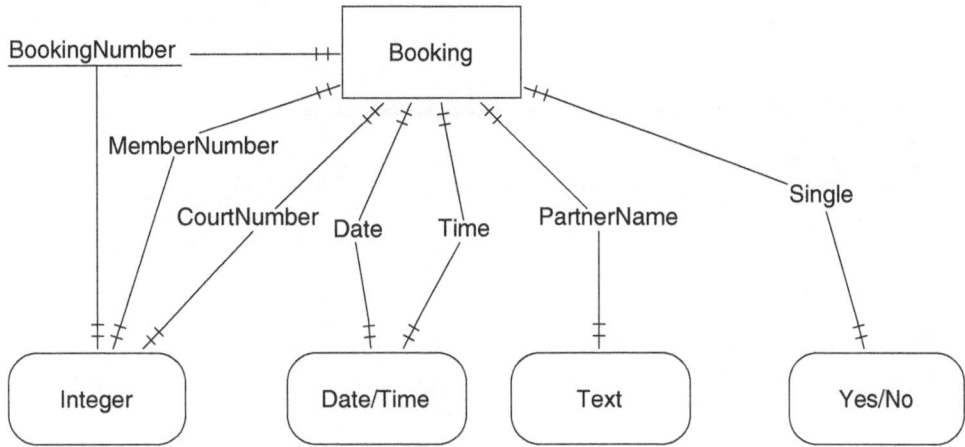

Figure 2.58 *Attribute diagram: Booking*

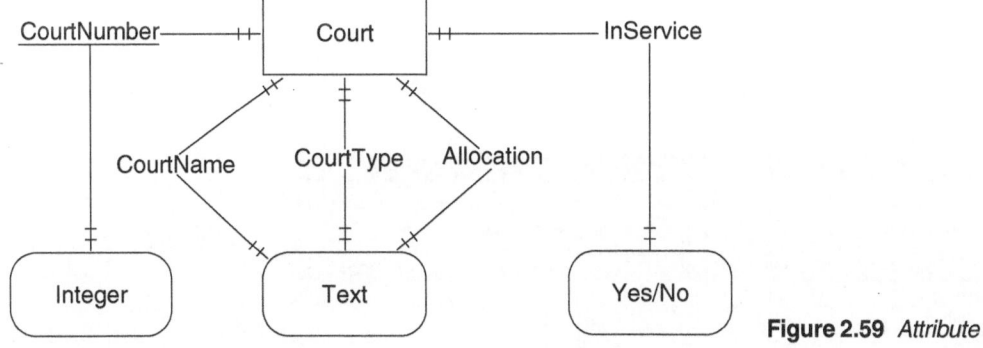

Figure 2.59 *Attribute diagram: Court*

(iii) Database tables

For the Member table it is best to create about 25 records to make it more realistic for applying the filter.

Remember the club has 10 courts. They can be grass, sand or hard courts.

Note that the relationships (and their associated degrees) defined in the design need to be implemented within the database application.

I have included 12 bookings in the Booking table.

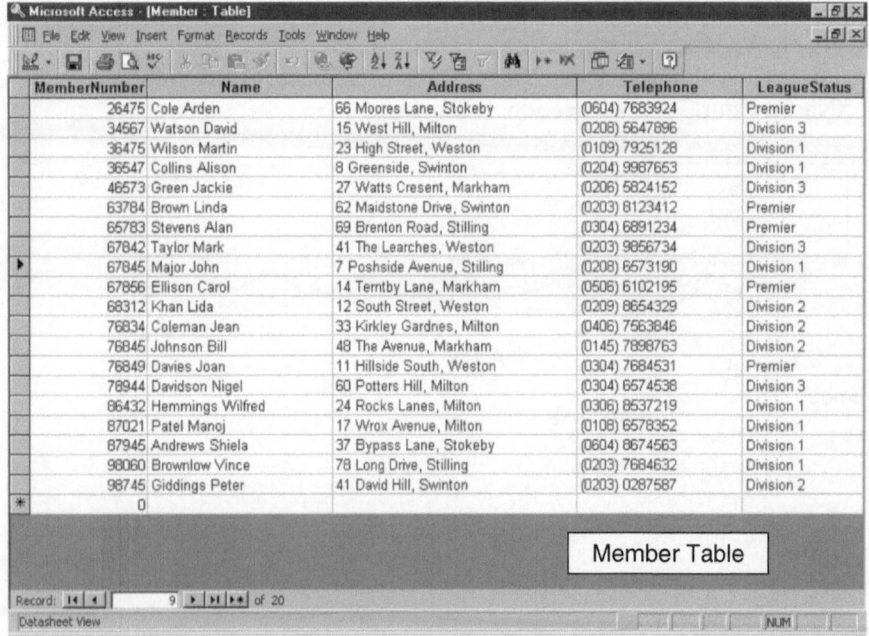

Member Table

Figure 2.60

Booking table

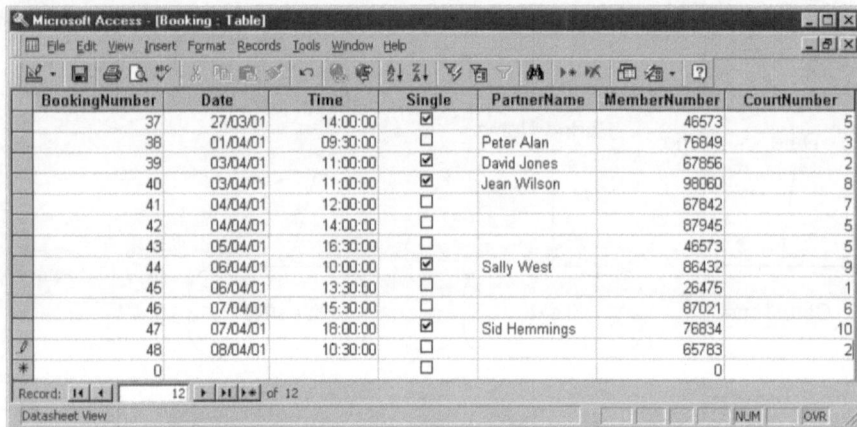

Figure 2.61

Court table

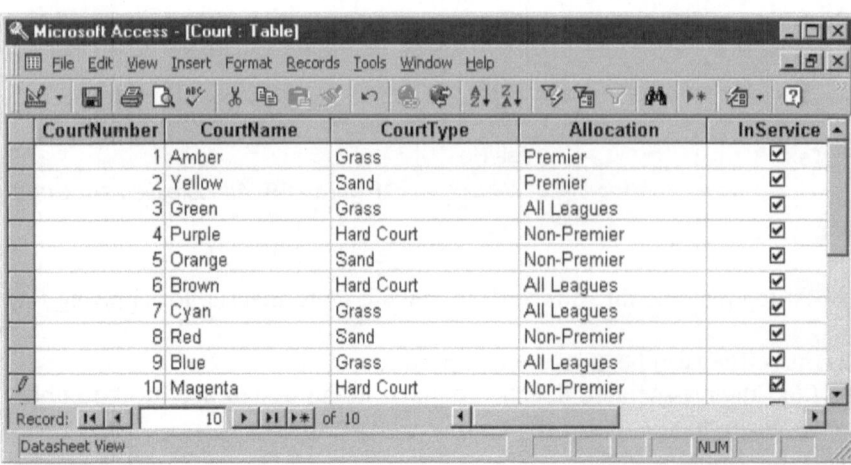

Figure 2.62

(iv) Member names in ascending order

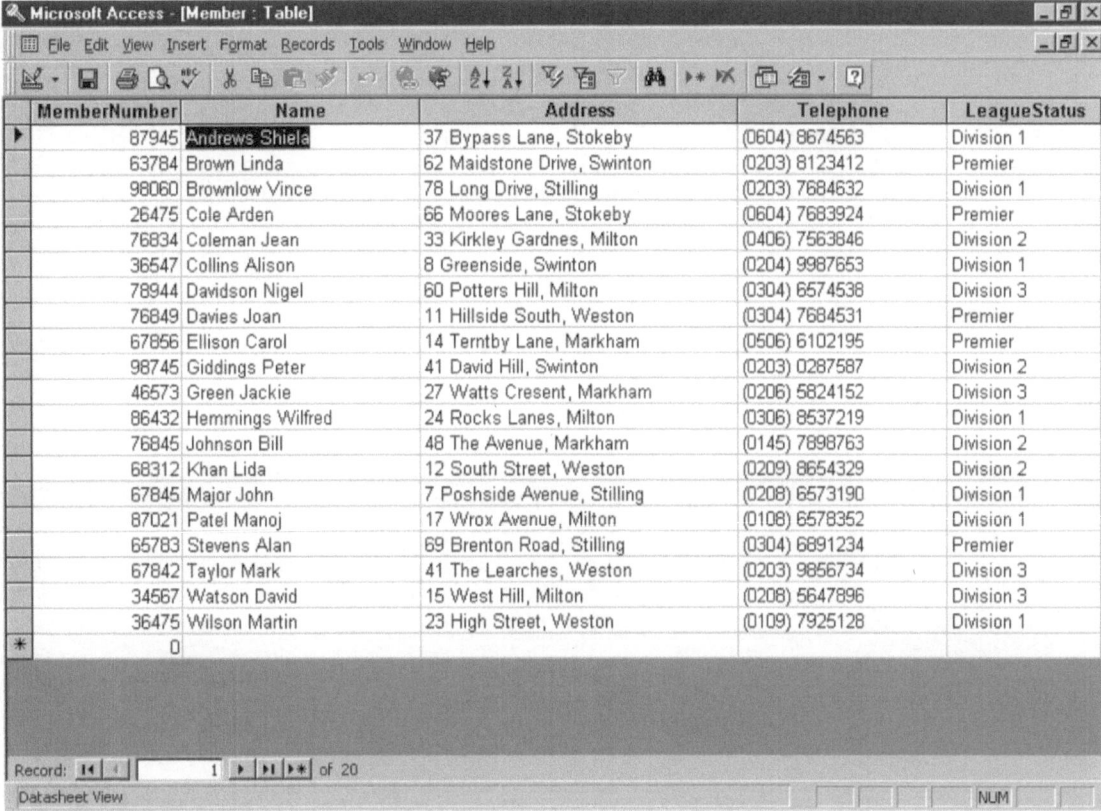

Figure 2.63

(v) Applying a filter (Selecting: Records, Filter, Advanced Filter/Sort)

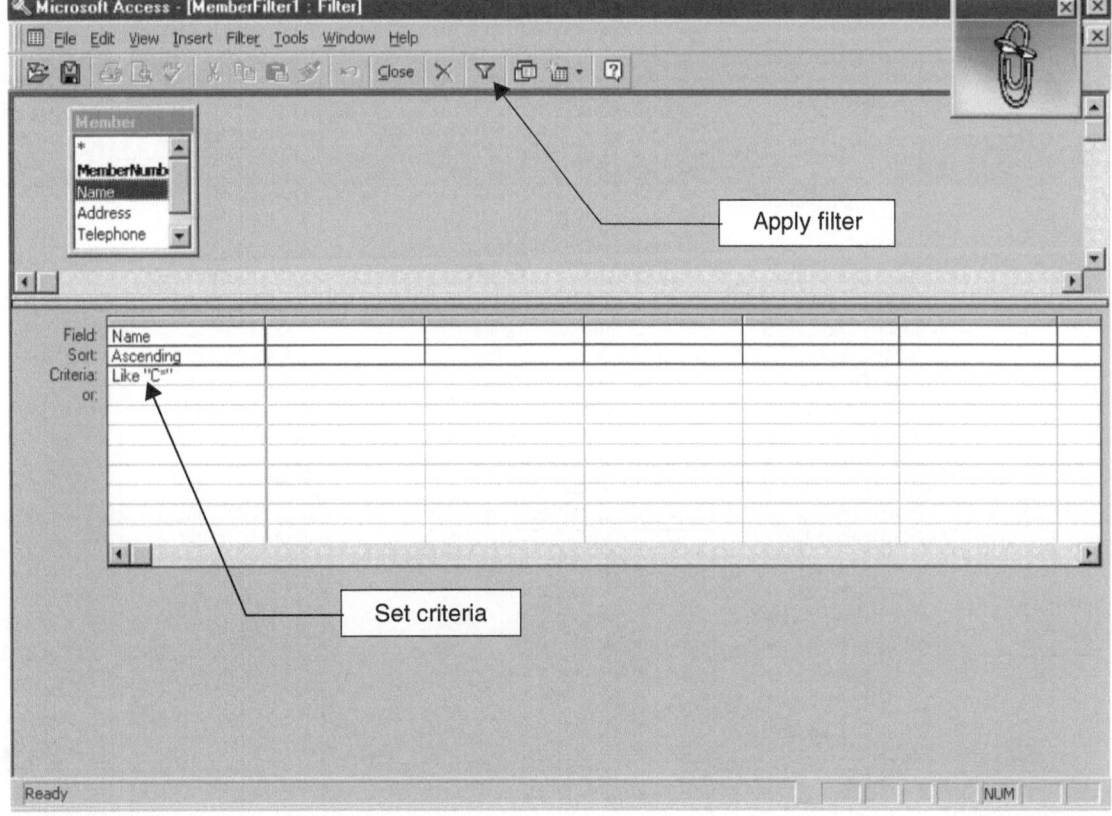

Figure 2.64

Produces

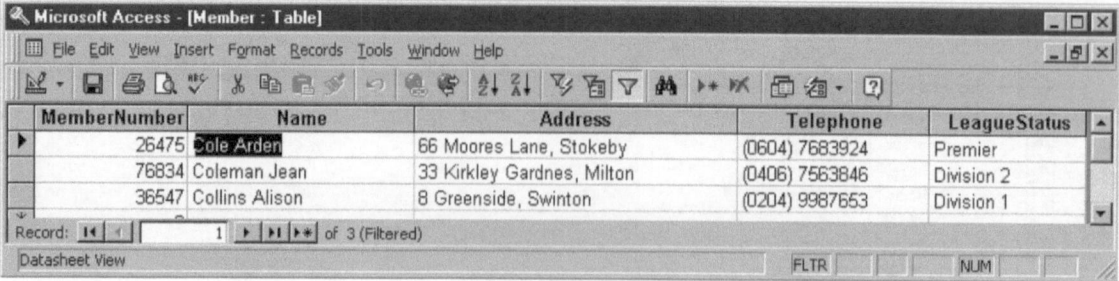

Figure 2.65

(vi)   Form view

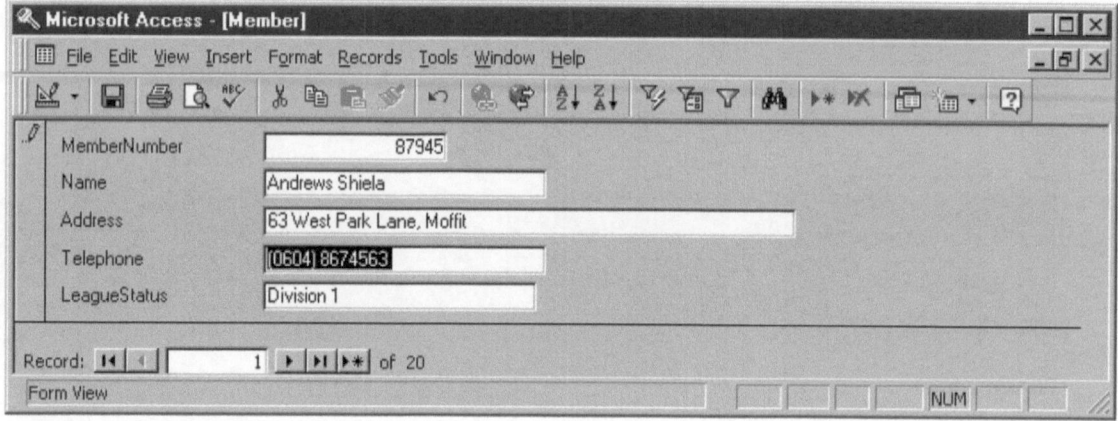

Figure 2.66

(vii) New record for Jill Milton added (also note the changes made in (vi))

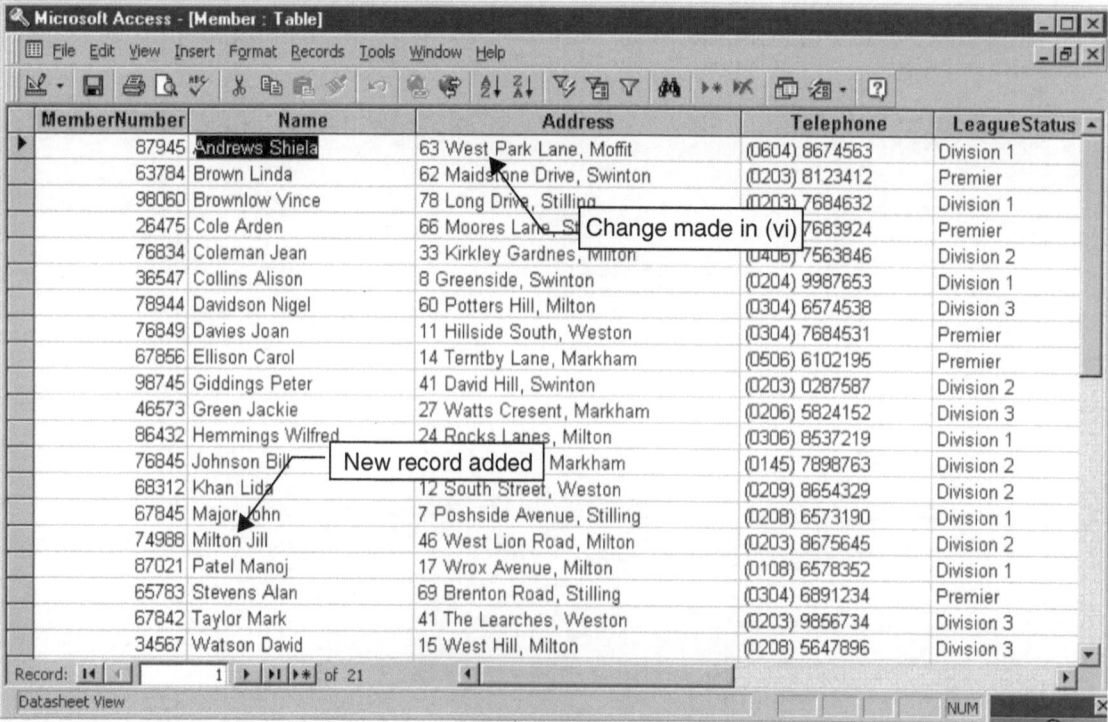

Figure 2.67

(viii) Establish relationships

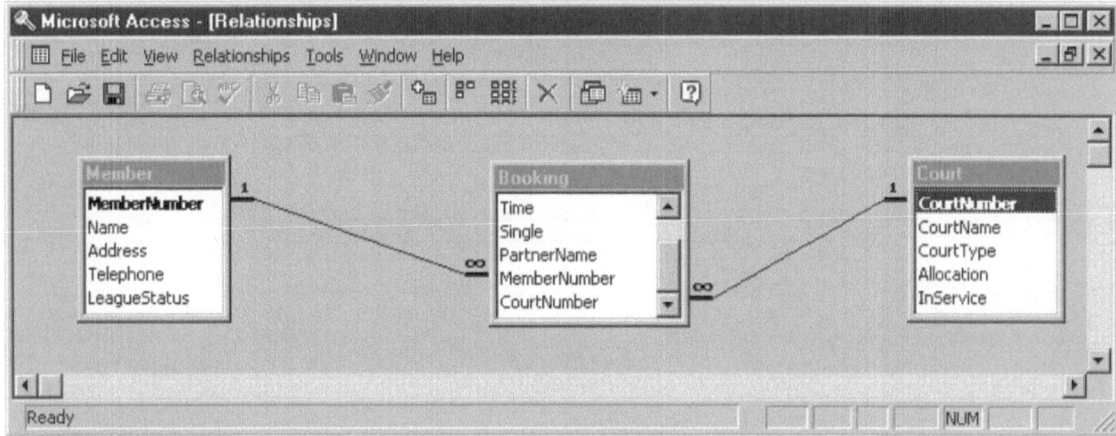

Figure 2.68

(ix) Subset Names v. League Status (select the Query tab from the main page)

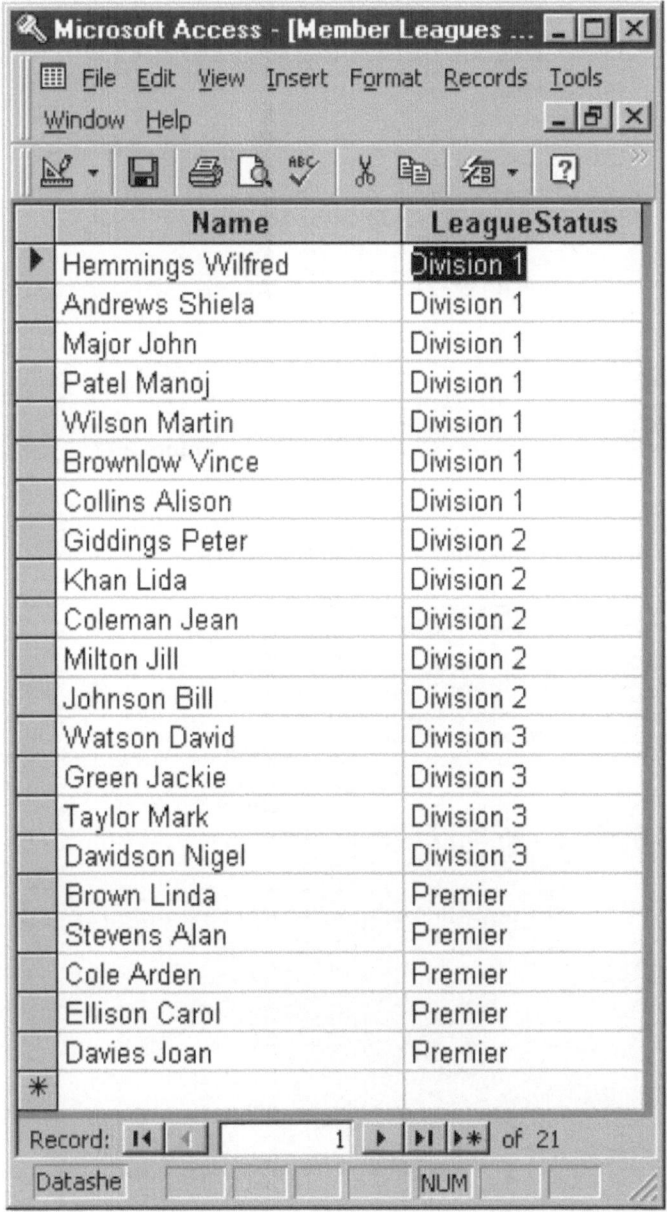

Name	LeagueStatus
Hemmings Wilfred	Division 1
Andrews Shiela	Division 1
Major John	Division 1
Patel Manoj	Division 1
Wilson Martin	Division 1
Brownlow Vince	Division 1
Collins Alison	Division 1
Giddings Peter	Division 2
Khan Lida	Division 2
Coleman Jean	Division 2
Milton Jill	Division 2
Johnson Bill	Division 2
Watson David	Division 3
Green Jackie	Division 3
Taylor Mark	Division 3
Davidson Nigel	Division 3
Brown Linda	Premier
Stevens Alan	Premier
Cole Arden	Premier
Ellison Carol	Premier
Davies Joan	Premier

Figure 2.69

(x)   Reports – Shown in Figure 2.70 is the Member Report
      Note: It may be necessary to adjust the design so the headings line up.

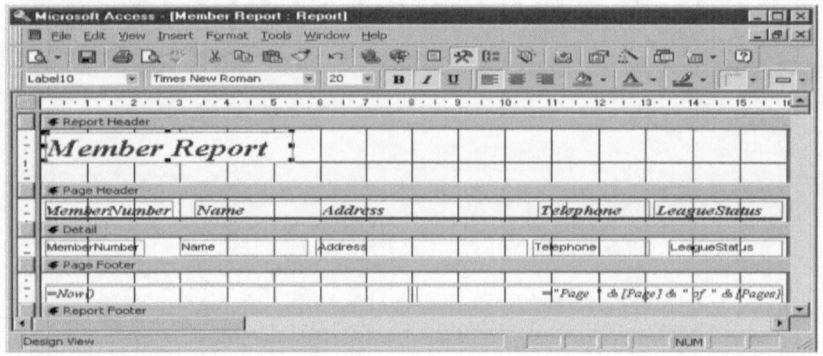

Figure 2.70

Member report

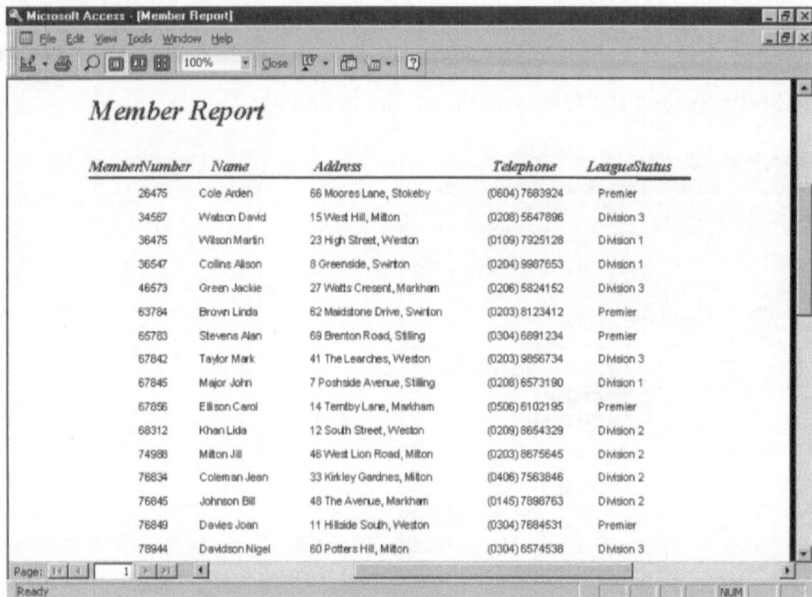

Figure 2.71

(xi)   Macro (outlined in Figure 2.72 is the design for the Macro – called Macro1)

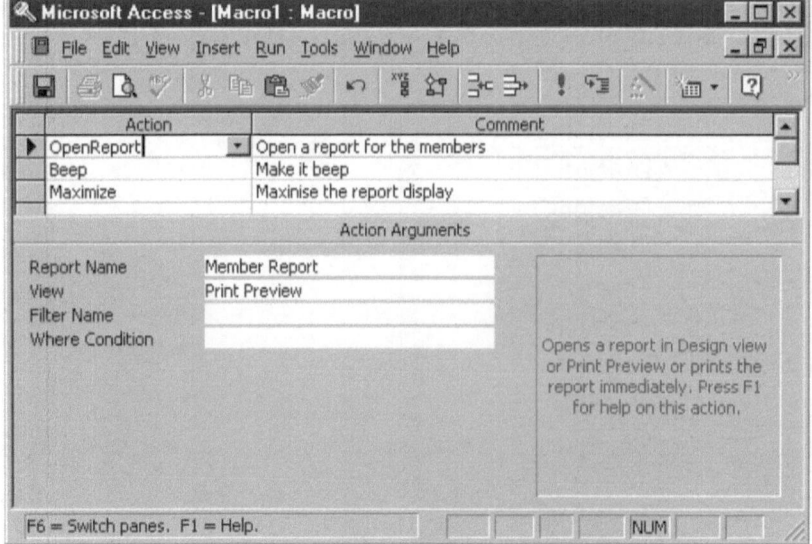

Figure 2.72

Programming concepts

Worksheet   Unit 3.1 Pascal

Fault finding in Pascal programs

The following Pascal program has at least one syntax or logical error on *each line*.
Identify each one.

```
program worksht one (input;output);

var;    x,y,z:integer;
        earnings,tax,x:real;
        st,for,name:string

begin
        write("What is your name ");
        readln(yourname);
        writeln('Hello ',name, ", how much to do earn each year? );
        readline(earnings);
        tax=earning*0.22;
        writeline(You will pay', tax:0:3,' pounds tax');

end;
```

Write your answers as Pascal comments near each fault.

Worksheet   Unit 3.2 Pascal

Fault finding in Pascal programs

The following Pascal program has at least five syntax or logical errors.
Identify each one.

```
program 2ndversion_of_table;
var counter:integer;
      i:real;

function getnumber(lower,upper:integer):integer;
var inputstr:string;
      errorcode:integer;
      x:integer;

begin
      while
            write('Type a value between ',lower,' and ', upper,' inclusive');
            readln(inputstr);
            val(inputstr,x,errorcode);
      until ((errorcode=0) and (x>=lower) and (x<=upper));

end;

begin {start of main program block }

      counter:=getnumber(2,12);
      for i:=1 to 12 do;
                  writeln(i,' times ',counter,' = ',i*counter);
end.
```

Write your answers as Pascal comments near each fault.

Worksheet   Unit 3.3 Pascal

Fault finding in Pascal programs

The following Pascal program has many syntax or logical errors.
Identify each one.

```
program area;

var a,b:real;
        x,y:float;

procedure area(a,b:real);
begin
        var area,a,b:real;
        area=a*b;
end.

begin
        print("Program to print area of rectangle");
        print("What is the length of side A? ");
        readln(a);
        print("What is the length of side B? );
        readln(b);
        area(a,b);
        print(The area is ',area:0:2);
end;
```

Write your answers as Pascal comments near each fault.

Worksheet   Unit 3.4 Pascal

Fault finding in Pascal programs
The following Pascal program has many syntax or logical errors.
Identify each one.

```
program conversion(output,input);
uses crt;

function miles2km(x:integer);
start
        milestokm:=x/1.609344;
end;

begin
        write(Program to convert miles to kilometres');
        write('What is the distance in miles? ')
        readln(miles);
        writeln(miles,' equals ',miles2km(miles),' kilometres);
        clrscr;
end.
```

Write your answers as Pascal comments near each fault.

Worksheet   Unit 3.5 Pascal

Find the error by using appropriate *test data*.

```
program prog4_23;

function titlecase(name:string):string;
{ returns string with first letter=capitals }

var     p,i:integer;
            ch:string;
            done:boolean;
            convert_to_caps:boolean;
            outputstr:string;

begin

        name:=lowercase(name); { first ensure everything is lowercase }

        done:=false;
        repeat {a loop to ensure that any leading spaces are removed }

                if copy(name,1,1)=' ' { if first char is a space }
                        then
                                { get rest of string only }
                                name:=copy(name,2,length(name)-1)
                        else
                                { assign boolean to end the loop }
                                done:=true;
        until done;

        convert_to_caps:=true; { initialise boolean for use in for loop }
        outputstr:=''; { initialise string to null
                                ready to build output string}

        for i:=1 to length(name) do; {look at each character in string }
                begin
                        ch:=copy(name,i,1); { get one char at a time }
                        if convert_to_caps=true
                                then
                                begin
                                        { add upper case char to output string }
                                        outputstr:=concat(outputstr,upcase(ch));
                                        { set boolean ready for next time}
                                        convert_to_caps:=false;
                                end
                                else
                                begin
                                        { add lower case char to output string }
                                        outputstr:=concat(outputstr,ch);
                                end;

                        { see if next char must be capital }
                        if ch=' ' then convert_to_caps:=true;
                end;
        titlecase:=outputstr; { assign finished string to the function }
end;

begin {start of main program block }
        {simply to test function output}
writeln(titlecase('this is a test string'));
end.
```

Worksheet   Unit 3.6 Pascal

Paint estimator 1

Write a program that estimates the amount of paint required to decorate the walls of a plain rectangular room. Your program should ask the user for the room's dimensions and the paint coverage per litre, then calculate the amount of paint required for a single coat of paint.

- Allow a yes/no question to ask if the ceiling is to be included in the calculation
- Assume a plain rectangular room
- Make no allowance for windows or doors
- Do not use any data input validation or error checking
- For a room with painted walls, emulsion paint covers $X \, m^2$ per litre. Use a coverage of $15 \, m^2$ per litre for program testing but your program should allow any value.

The calculation is ((area of wall A $\times$ 2) + (area of wall B $\times$ 2))/paint coverage. If the ceiling is to be painted, use the wall dimensions to calculate ceiling size then round up to the nearest litre. Only round up total wall + ceiling paint required at the end of the program to avoid two rounding errors.

Points to note

- Use Stepwise Refinement to design your program
- *Before* coding, design simple test data
- Use sensible variable and procedure names
- Use appropriate variable types
- When rounding up the paint required, allow for a small error in coverage, i.e. if the amount of paint is only 2% or less above an integer number of litres, round it down. For example, suppose the paint required were 7.03 litres – 2% of 7 litres = $7 \times 2/100 = 0.14$. As 0.03 is below 0.14, round it down. Putting it another way, 7.14/7 = 1.02 meaning 7.14 is 2% above 7.

Worksheet   Unit 3.7 Pascal

Paint estimator 2

Extend your paint estimator program to calculate the paint costs.

A local paint shop sells paint with these prices. Extend the program to ask if white or a colour is required then output the paint cost. Only allow for the ceiling to be the same colour as the walls.

Size	Brilliant White	Colours
10 litre	£26.99	£39.99
5 litre	£13.49	£20.99
2½ litre	£9.29	£12.99
1 litre	£4.99	£6.49

You should use some integer arithmetic DIV and MOD to arrive at the correct number of tins to obtain the best price. For example, if a job required 12 litres of coloured paint, you could make 12 litres by

12 by 1 litre tins =	£77.88
2 by 5 litre tins plus 2 by 1 litre tins =	£54.96
4 by 2½ tins plus 2 by 1 litre tins =	£64.94

Clearly, the best price is £54.96.

Worksheet   Unit 3.8 Pascal

Paint estimator 3

Rewrite the paint estimator program to use procedures.

Write a brief answer to the questions:

1. What advantage does the use of procedures give over a single piece of program code regarding:
    * The user?
    * The programmer?
    * The programmer's employer?
2. What disadvantage occurs when using only global variables?

Answers to worksheets

Worksheet Unit 3.1 Pascal

```
program worksht one (input;output);
        {space in program name, ; instead of , in input,output }

var;    x,y,z:integer;
        { ; instead of : after var }

        earnings,tax,x:real;
        { x defined as variable but name x has been used already}

        st,for,name:string
        {no ; at the end of the statement}
        {reserved word "for" used as var name }

begin
        write("What is your name ");
        { wrong quote marks, should be ' ' }

        readln(yourname);
        { variable not defined }

        writeln('Hello ',name, ', how much to do earn each year? );
        { missing quote after year? }

        readline(earnings);
        { should be readln not readline }

        tax=earning*0.22;
        { variable earning not defined, should be earnings }

        writeline(You will pay', tax:0:3,' pounds tax');
        { writeline should be writeln }

end;    { end of program should be end. not end; }
```

Worksheet Unit 3.2 Pascal

```
program 2ndversion_of_table;
        { cannot start an identifier with a number (2) }
var counter:integer;
        i:real;

function getnumber(lower,upper:integer):integer;
var inputstr:string;
        errorcode:integer;
        x:integer;

begin
        while
        {while should be repeat }
                write('Type a value between ',lower,' and ',upper,' inclusive');
                readln(inputstr);
                val(inputstr,x,errorcode);
        until ((errorcode=0) and (x>=lower) and (x<=upper));
        { function value must be assigned to function name with getnumber:=x; )
end;

begin {start of main program block }

        counter:=getnumber(2,12);
        for i:=1 to 12 do;
                {i is of var type real so cannot be used as a counter here }
                { the ; after do will stop the loop executing }
                        writeln(i,' times ',counter,' = ',i*counter);
end.
```

Worksheet   Unit 3.3 Pascal

```
program area;

var a,b:real;
      x,y:float;
              { float is not a Pascal variable type, it belongs in c }

procedure area(a,b:real);
              {identifier "area" has been used already as the program name,
                   identifiers must be unique }
begin
      var area,a,b:real;
              { var definition block must be outside execution block }
              { identifier "area" has been used already,
                   identifiers must be unique }
      area=a*b;
              { assignment operator should be := }
end.
              { end of procedure definition should be end; }


begin
      print("Program to print area of rectangle");
              { print is a BASIC reserved word, should be write }
              { wrong quotes, should be ' ' }
      print("What is the length of side A? ");
              { print is a BASIC reserved word, should be write }
              { wrong quotes, should be ' ' }
      readln(a);
      print('What is the length of side B? );
              { print is a BASIC reserved word, should be write }
              { missing quote after B? }
      readln(b);
              { should be readln not readln}
      area(a,b);
      print(The area is ',area:0:2);
              { print is a BASIC reserved word, should be write }
              { no way for a value to be returned, variable out of scope }
              { missing quote before The }
end;
              { end of program definition should be end. }
```

Worksheet   Unit 3.4 Pascal

```
program conversion(output,input);
      { In standard Pascal, the program header ends in (input, output) }
      { (input,output) optional in Free Pascal and Borland Turbo Pascal }

uses crt;

function miles2km(x:integer);
      { no function return type, should be real }
start
      { should be begin not start }
      milestokm:=x/1.609344;
      { should be x*1.609344 not x/1.609344 }
      { user defined function name is miles2km not milestokm }
end;

begin
      write(Program to convert miles to kilometres');
      { missing quote at start of string }
      write('What is the distance in miles? ')
      { missing ; character at end of statement }
      readln(miles);
      { var not defined }
      writeln(miles,' equals ',miles2 km(miles),' kilometres);
      { missing quote after kilometres }
      clrscr;
      { cause a blank screen so output will not be visible }
end.
```

Worksheet   Unit 3.5 Pascal

The statement `for i:=1 to length(name) do;` has a ';' indicating end of statement but the real end of statement is the 'end;' directly below the 'begin' of the 'for' loop. The result is that no looping will occur. Pascal will count from 1 to the length of the string then execute what should have been looped but only once. The program with the fault in place will output a single character of the string only (in capitals) as the variable 'i' will not have been assigned an initial value.

```
for i:=1 to length(name) do; { ; should not be here }
      begin
            ch:=copy(name,i,1); { get one char at a time }
            if convert_to_caps=true
                  then
                  begin
                        { add upper case char to output string }
                        outputstr:=concat(outputstr,upcase(ch));
                        { set boolean ready for next time}
                        convert_to_caps:=false;
                  end
                  else
                  begin
                        { add lower case char to output string }
                        outputstr:=concat(outputstr,ch);
                  end;

            { see if next char must be capital }
            if ch=' ' then convert_to_caps:=true;
      end; { end of "for" loop }
```

Worksheet   Unit 3.6 Pascal to Unit 3.8 Pascal

Using Stepwise Refinement, break down the problem into these parts:

- Get data, i.e. get room dimensions and paint coverage
- Do calculations
- Output results.

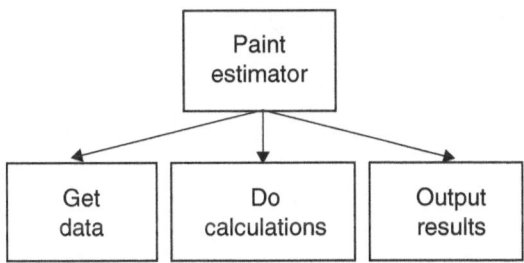

Figure 3.1

Example test data without data validation tests

Width of wall A	Width of wall B	Height	Paint coverage	Ceiling area	Wall area	Total area	Litres to paint walls only	Litres to paint all	Rounded litres to paint walls only	Rounded litres to paint all
3	2	2.5	15	6	25	31	1.67	2.07	2	3
5	3	3	15	15	48	63	3.20	4.20	4	5
1	1.5	2	15	1.5	10	11.5	0.67	0.77	1	1
1.5	4	2.7	17	6	30	35.7	1.75	2.10	2	3
2.9	2	2.5	15	5.8	25	30.3	1.63	2.02	2	2

Test sheet

Inputs					Rounded litres		OK
Width of wall A	Width of wall B	Height	Paint coverage	Paint ceiling?	Expected output	Actual output	Y or N
3	2	2.5	15	Y	3		
5	3	3	15	Y	5		
1	1.5	2	15	Y	1		
1.5	4	2.7	17	Y	3		
2.9	2	2.5	15	Y	2		
3	2	2.5	15	N	2		
5	3	3	15	N	4		
1	1.5	2	15	N	1		
1.5	4	2.7	17	N	2		
2.9	2	2.5	15	N	2		

First attempt at program (with a bug as will be shown below)

```
program trp101;

{Program instructions included here }
{The calculation is
((area of wall A*2) + (area of wall B*2)) / paint coverage.
If the ceiling is to be painted, use the wall dimensions to
calculate ceiling size then round up to the nearest litre.
Only round up total wall+ceiling paint required at the end
of the program to avoid 2 rounding errors.}

var wallA,wallB,height,coverage:real;
         wallarea,totalarea,paintneeded:real;
     paintceiling:boolean;
     answer:char;

begin
            {start of getdata section}
            writeln('Paint coverage program version 1.0');
            writeln('Please input values in metres');
            write('What is the length of the first wall? ');
            readln(wallA);
            write('What is the length of the second wall? ');
            readln(wallB);
            write('What is the height of the room? ');
            readln(height);
            write('Do you wish to allow for painting the ceiling,
                 type Y or N? ');
            readln(answer);
            write('What is the coverage of the paint
                 in square metres per litre? ');
            readln(coverage);

            {start of calculation section }
            if (answer='Y') or (answer='y') then
                 {allow any other key to be equal to "no"}

                 paintceiling:=true
                 else
                 paintceiling:=false;

            wallarea:=(wallA*height*2)+(wallB*height*2);

            if paintceiling then
                 totalarea:=wallarea+(wallA*wallB)
                 else
```

```
                     totalarea:=wallarea;

              paintneeded:=totalarea/coverage;

              {check for 2% tolerance in paint quantity}
                     if paintneeded/int(paintneeded) >=1.02 then
                         paintneeded:=int(paintneeded)+1
                         else
                         paintneeded:=int(paintneeded);
              {start of output section}
              write('Paint required is ',paintneeded:0:0,' litres');
              if paintceiling then
                     write(' to cover walls and ceiling')
                     else
                     write(' to cover the walls only');
end.
```

Output using data from line 1 of the test data

```
Paint coverage program version 1.0
Please input values in metres
What is the length of the first wall? 3
What is the length of the second wall? 2
What is the height of the room? 2.5
Do you wish to allow for painting the ceiling, type Y or N? y
What is the coverage of the paint in square metres per litre? 15
Paint required is 3 litres to cover walls and ceiling
```

From this data the program seems to work but if all the lines in the test data are used, a problem arises as is shown in the test data table.

Inputs					Rounded litres		OK
Width of wall A	Width of wall B	Height	Paint coverage	Paint ceiling?	Expected output	Actual output	Y or N
3	2	2.5	15	Y	3	3	Y
5	3	3	15	Y	5	5	Y
1	1.5	2	15	Y	1	Div by 0	N
1.5	4	2.7	17	Y	3	3	Y
2.9	2	2.5	15	Y	2	2	Y
3	2	2.5	15	N	2	2	Y
5	3	3	15	N	4	4	Y
1	1.5	2	15	N	1	Div by 0	N
1.5	4	2.7	17	N	2	2	Y
2.9	2	2.5	15	N	2	2	Y

Using data from the 3rd and 8th lines of the test data causes a program crash with 'Divide by zero'. There are only two division operations in the program so the bug must be from one of these lines:

```
paintneeded:=totalarea/coverage;
or
if paintneeded/int(paintneeded) >=1.02 then
```

The first one cannot generate a divide by 0 error if the value of the variable coverage is not zero, the bug must be in the second division. This line is used in the rounding decision, i.e. if the amount is within 2% of an integer number of litres. In the case where the test data gave a divide by 0 error, the value of int(paintneeded) is 0 as the amount of paint required is small. The solution is to test the value of int(paintneeded) before the division like this:

```
if int (paintneeded) >0 then
       if paintneeded/int(paintneeded) >=1.02 then
              paintneeded:=int(paintneeded)+1
              else
              paintneeded:=int(paintneeded);
if int(paintneeded)=0 then paintneeded:=1;
```

So the complete program is now

```
program trp101a;

{Program instructions included here }
{The calculation is
((area of wall A*2) + (area of wall B*2)) / paint coverage.
If the ceiling is to be painted, use the wall dimensions to
calculate ceiling size then round up to the nearest litre.
Only round up total wall+ceiling paint required at the end
of the program to avoid 2 rounding errors.}

var wallA,wallB,height,coverage:real;
    wallarea,totalarea,paintneeded:real;
    paintceiling:boolean;
    answer:char;

begin
    {start of getdata section}
    writeln('Paint coverage program version 1.0');
    writeln('Please input values in metres');
    write('What is the length of the first wall? ');
    readln(wallA);
    write('What is the length of the second wall? ');
    readln(wallB);
    write('What is the height of the room? ');
    readln(height);
    write('Do you wish to allow for painting the ceiling,
          type Y or N? ');
    readln(answer);
    write('What is the coverage of the paint
          in square metres per litre? ');
    readln(coverage);

    {start of calculation section }
        if (answer='Y') or (answer='y') then
            {allow any other key to be equal to "no"}

                paintceiling:=true
                else
                paintceiling:=false;

    wallarea:=(wallA*height*2)+(wallB*height*2);

    if paintceiling then
                totalarea:=wallarea+(wallA*wallB)
                else
                totalarea:=wallarea;

    paintneeded:=totalarea/coverage;

    {check for 2% tolerance in paint quantity}

        if int(paintneeded)>0 then {line added as result of testing}

        if paintneeded/int(paintneeded) >=1.02 then
            paintneeded:=int(paintneeded)+1
            else
            paintneeded:=int(paintneeded);

    if int(paintneeded)=0 then paintneeded:=1;
            {line added as result of testing}

    {start of output section}
    write('Paint required is ',paintneeded:0:0,' litres');
    if paintceiling then
                write(' to cover walls and ceiling')
                else
                write(' to cover the walls only');
end.
```

Worksheet   Unit 3.7 Pascal
Test sheet

	Expected outputs				Actual outputs				OK Y/N
White	10	5	1	total £	10	5	1	total £	
1	0	0	1	4.99	0	0	1	4.99	Y
2	0	0	2	9.98	0	0	2	9.98	Y
3	0	0	3	14.97	0	0	3	14.97	Y
4	0	0	4	19.96	0	0	4	19.96	Y
5	0	1	0	13.49	0	1	0	13.49	Y
6	0	1	1	18.48	0	1	1	18.48	Y
7	0	1	2	23.47	0	1	2	23.47	Y
8	0	1	3	28.46	0	1	3	28.46	Y
9	0	1	4	33.45	0	1	4	33.45	Y
10	1	0	0	26.99	1	0	0	26.99	Y
15	1	1	0	40.48	1	1	0	40.48	Y
16	1	1	1	45.47	1	1	1	45.47	Y
21	2	0	1	58.97	2	0	1	58.97	Y
23	2	0	3	68.95	2	0	3	68.95	Y
50	3	0	0	80.97	3	0	0	80.97	Y
Colours	10	5	1		10	5	1		
1	0	0	1	6.49	0	0	1	6.49	Y
2	0	0	2	12.98	0	0	2	12.98	Y
3	0	0	3	19.47	0	0	3	19.47	Y
4	0	0	4	25.96	0	0	4	25.96	Y
5	0	1	0	20.99	0	1	0	20.99	Y
6	0	1	1	27.48	0	1	1	27.48	Y
7	0	1	2	33.97	0	1	2	33.97	Y
8	0	1	3	40.46	0	1	3	40.46	Y
9	0	1	4	46.95	0	1	4	46.95	Y
10	1	0	0	39.99	1	0	0	39.99	Y
15	1	1	0	60.98	1	1	0	60.98	Y
16	1	1	1	67.47	1	1	1	67.47	Y
21	2	0	1	86.47	2	0	1	86.47	Y
23	2	0	3	99.45	2	0	3	99.45	Y
50	3	0	0	119.97	3	0	0	119.97	Y

```
program trp102;

{As program TRP101a but with addition of paint costs}

var wallA,wallB,height,coverage:real;
        wallarea,totalarea,paintneeded:real;
      paintceiling:boolean;
      answer:char;
      {new variables needed}
      intpaintneeded:integer;
      size10tins,size5tins,size1tins:integer;
      cost:real;
      iswhite:boolean;
      whatcolour:char;

begin
              {start of getdata section}
              writeln('Paint coverage program version 1.0');
              writeln('Please input values in metres');
              write('What is the length of the first wall? ');
              readln(wallA);
              write('What is the length of the second wall? ');
              readln(wallB);
              write('What is the height of the room? ');
              readln(height);
              write('Do you wish to allow for painting the ceiling,
                    type Y or N? ');
              readln(answer);
              write('What is the coverage of the paint in square
                    metres per litre? ');
              readln(coverage);
              write('What colour is required, W for white, C for colour ');
              readln(whatcolour);

              {start of calculation section }

              {allow any other key to be equal to "no"}
              if (answer='Y') or (answer='y') then paintceiling:=true
                    else
                    paintceiling:=false;

              wallarea:=(wallA*height*2)+(wallB*height*2);

              if paintceiling then
                    totalarea:=wallarea+(wallA*wallB)
                    else
                    totalarea:=wallarea;

              paintneeded:=totalarea/coverage;

              {check for 2% tolerance in paint quantity}

      if int (paintneeded)>0 then
                    if paintneeded/int(paintneeded) >=1.02 then
                          paintneeded:=int(paintneeded)+1
                          else
                          paintneeded:=int(paintneeded);

      if int(paintneeded)=0 then paintneeded:=1;
                    {line added as result of testing to cope with small amounts}

      {now calculate costs}

      intpaintneeded:=trunc(paintneeded);
              {the trunc function will convert the real type paintneeded to
              an integer type intpaintneeded. In Pascal, you must be very
              careful with variable types. }
```

```
{use of integer arithmetic DIV and MOD to calculate number of tins}
{ DIV will give the number of tins and MOD gives the remainder
after using DIV ready to find the number of the next smaller size }

size10tins:= intpaintneeded DIV 10;
intpaintneeded:= intpaintneeded MOD 10;

size5tins:= intpaintneeded DIV 5;
intpaintneeded:= intpaintneeded MOD 5;

size1tins:= intpaintneeded;

        if (whatcolour='W') or (whatcolour='w') then
                    iswhite:=true
            else
                    iswhite:=false;

    if iswhite
        then
        cost:=(size10tins*26.99)+(size5tins*13.49)
        +(size1tins*4.99)

    else
        cost:=(size10tins*39.99)+(size5tins*20.99)
        +(size1tins*6.49);

    {start of output section}
    write('Paint required is ',paintneeded:0:0,' litres');
    if paintceiling then
        write(' to cover walls and ceiling')
        else
        write(' to cover the walls only');

    writeln;
    if size10tins>0 then {to avoid writing 0 tins which looks ugly}
        writeln('You will need ',size10tins,' 10 litre tins');

    if size5tins>0 then
        writeln('You will need ',size5tins,' 5 litre tins');

    if size1tins>0 then
        writeln('You will need ',size1tins,' 1 litre tins');

        writeln('The total cost will be ',cost:0:2,' pounds');

end.
```

Worksheet   Unit 3.8 Pascal

1. What advantage does the use of procedures give over a single piece of program code regarding:
   - The user?
   None except that the program is more likely to be free of errors as a result of better program design and coding.
   - The programmer?
   For programs larger than a few lines, the use of procedures enables the programmer to think more clearly. If the overall programming task is large, each part may be written and tested in isolation. It is also easier to translate the program design into procedures as most design techniques assume that code is split into manageable parts.
   - The programmer's employer?
   For larger programs, the management can ensure the programming team have specialists that may be asked to complete certain parts of the project. For example, a specialist in screen layout could design the screen and a specialist in accounting would ensure the costings are accurate.
2. What disadvantage occurs when using only global variables?
   Especially with reference to the answers above, when programs are written by teams of people, it is clearly impractical to ensure they each have their own set of global variable names to ensure one section does not clash with another. Local variable ensures that the value in a local variable will not cause unforeseen problems in other sections of the whole program.

The following program is the same as trp102 except that the program is split into procedures. These have a direct relationship with the structure as described in the program design using Stepwise Refinement.

Only global variables have been used.

```
program trp103;

{As program TRP102 but with procedures}

var wallA,wallB,height,coverage:real;
        wallarea,totalarea,paintneeded:real;
    paintceiling:boolean;
    answer:char;
    {new variables needed}
    intpaintneeded:integer;
    size10tins,size5tins,size1tins:integer;
    cost:real;
    iswhite:boolean;
    whatcolour:char;

procedure getdata;
begin
    {start of getdata section}
    writeln('Paint coverage program version 1.0');
    writeln('Please input values in metres');
    write('What is the length of the first wall? ');
    readln(wallA);
    write('What is the length of the second wall? ');
    readln(wallB);
    write('What is the height of the room? ');
    readln(height);
    write('Do you wish to allow for painting the ceiling,
            type Y or N? ');
    readln(answer);
    write('What is the coverage of the paint in
            square metres per litre? ');
    readln(coverage);
    write('What colour is required, W for white, C for colour ');
    readln(whatcolour);
end;
```

. . . .

```
procedure calculate_paint_needed;

begin
    {allow any other key to be equal to "no"}
    if (answer='Y') or (answer='y') then paintceiling:=true
                else
                paintceiling:=false;

    wallarea:=(wallA*height*2)+(wallB*height*2);

    if paintceiling then
            totalarea:=wallarea+(wallA*wallB)
            else
            totalarea:=wallarea;

    paintneeded:=totalarea/coverage;

    {check for 2% tolerance in paint quantity}

    if int(paintneeded)>0 then
            if paintneeded/int(paintneeded) >=1.02 then
                        paintneeded:=int(paintneeded)+1
                        else
                        paintneeded:=int(paintneeded);

    if int(paintneeded)=0 then paintneeded:=1;
            {line added as result of testing to cope with small amounts}
end;
```

. . . .

```
procedure calc_cost;

        intpaintneeded:=trunc(paintneeded);
           {the trunc function will convert the real type paintneeded to
           an integer type intpaintneeded. In Pascal, you must be very
           careful with variable types. }

        {use of integer arithmetic DIV and MOD to calculate number of tins}
        { DIV will give the number of tins and MOD gives the remainder
        after using DIV ready to find the number of the next smaller size }

        size10tins := intpaintneeded DIV 10;
        intpaintneeded := intpaintneeded MOD 10;

        size5tins := intpaintneeded DIV 5;
        intpaintneeded := intpaintneeded MOD 5;

        size1tins := intpaintneeded;

                if (whatcolour='W') or (whatcolour='w') then
                            iswhite := true
                     else
                            iswhite := false;
                if iswhite
                     then
                     cost:=(size10tins*26.99)+(size5tins*13.49)
                     +(size1tins*4.99)
                else
                     cost:=(size10tins*39.99)+(size5tins*20.99)
                     +(size1tins*6.49);
end;

.......................................................................

procedure output_results;

        write('Paint required is ',paintneeded:0:0,' litres');
        if paintceiling then
           write(' to cover walls and ceiling')
           else
           write(' to cover the walls only');

        writeln;
        if size10tins>0 then {to avoid writing 0 tins which looks ugly}
           writeln('You will need ',size10tins,' 10 litre tins');

        if size5tins>0 then
           writeln('You will need ',size5tins,' 5 litre tins');

        if size1tins>0 then
           writeln('You will need ',size1tins,' 1 litre tins');

           writeln('The total cost will be ',cost:0:2,' pounds');

end;

.......................................................................

begin {start of main program }

        getdata;
        calculate_paint_needed;
        calc_cost;
        output_results;

end.
```

Concepts of database design

4.1 Activity One – File-based systems

File-based systems use application programs to define and manage data, modelled around the concept of decentralization.

In an organization each functional department generates its own data sets, this data would then be electronically stored, manipulated and accessed by personnel within that functional area as shown in Figure 4.1.

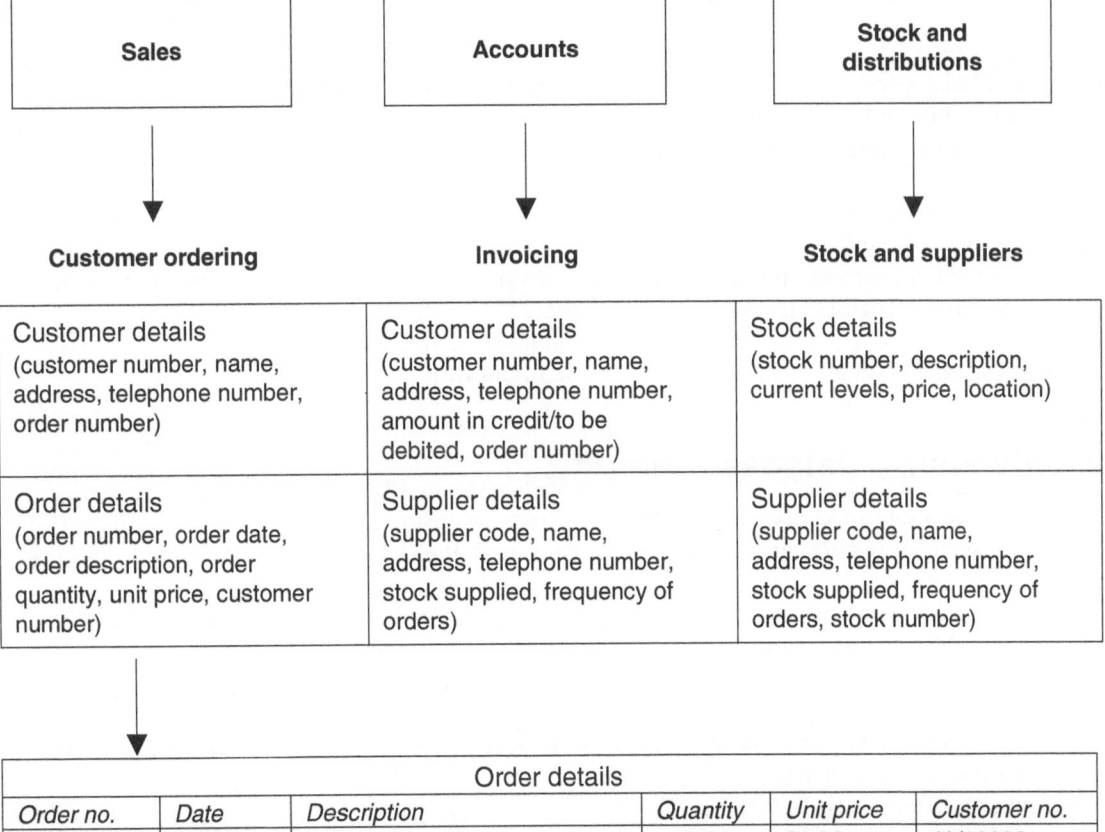

Order details					
Order no.	*Date*	*Description*	*Quantity*	*Unit price*	*Customer no.*
P123/4O	23/06/03	2 cm rubber rings	150	£0.24	JA/19009
R239/7G	24/06/03	2 m² wire mesh sheets	30	£1.12	GH/34899
P349/9I	24/06/03	10 cm plastic screw caps	100	£0.46	JA/19009
S754/8K	25/06/03	1 m copper pipes	40	£1.07	DR/47886
HY492/0P	25/06/03	1 m² insulating foam sheets	25	£0.89	DR/47886

Figure 4.1

Tasks
1. Based on the example given produce a similar representation of a file-based system for:
   (i) an employee record within a human resource department
   (ii) stock despatch details to a customer within a distributions department.
2. Describe in detail four problems with a decentralized approach to file management.

4.2 Activity Two – Data models

Databases have evolved from users and developers being able to understand the semantics of data sets and communicating this understanding clearly and logically. To facilitate this, a specific data model/s can be used as a framework for examining and understanding the entities, attributes and relationships between data sets.

Data models can be broken down into three categories, these include:

- Object-based models: entity-relationship, semantic, object-orientated and functional
- Record-based models: hierarchical, network and relational
- Physical data models.

Tasks

1. Provide a short summary outlining these three data model categories and the three elements that they consist of.
2. Compare and contrast the following data models:
   (i)   Hierarchical
   (ii) Network
   (iii) Relational.
3. Using diagrams, provide a visual representation of a hierarchical, network and relational data model.

4.3 Activity Three – Database Management Systems (DBMS)

Tasks

1. Provide a detailed overview of the components that contribute to a DBMS
2. Describe the benefits and limitations of a DBMS
3. Identify three software contenders that support DBMS and outline some of the main features and tools of each
4. Carry out research on an organization that has implemented a DBMS and produce a 1500 word report incorporating some of the following criteria:
   (i)   Costs incurred in terms of updating hardware, software, training and maintenance of the system
   (ii) What systems or processes the DBMS has integrated
   (iii) What provisions are in place for updating and maintaining the system
   (iv) What impact the DBMS has had/is having on the organization.

4.4 Activity Four – Database security

Security of data is essential especially with a database system, to ensure privacy of sensitive and personal information. Data security is also paramount in complying with legislation that protects users and third parties of data.

A number of organizational security breaches can occur, some of these are amplified by the use of a database because of the integrated approach to data storage and retrieval.

Tasks

1. For each of the following security breaches describe the measures that can be used to overcome them:
   (i)   Threat of hackers (external)
   (ii) Unauthorized user access (internal)
   (iii) Viruses
   (iv) Vandalism or sabotage.
2. What legislation can be enforced to combat potential security breaches and what are the consequences if a breach is made by an individual.

4.5 Activity Five – Database design and editing

Scenario

You have been asked to set up a database for your local surgery. The surgery has several functions that need to be computerized using suitable database software. These areas include:

- Doctors and locums – name, registration number, surgery appointment times, specialist areas and patient list

- Appointments – dates, doctors, patients, description/nature of the appointment (migraine, aches and pains etc.)
- Patients – hospital number, name, address, telephone number and date of birth
- Personnel – surgery staff (name, address, telephone number, date of birth, employee number, hourly pay rate, date of employment etc.).

The surgery staff consist of Jan Greene who is the surgery manager and three full-time administration staff, who schedule the appointments. There is one nurse and five doctors and locums also in practice at the surgery.

Brief

You have been asked to perform a number of tasks to include an overall design of the proposed database system, the development of a fully working design, reports, updating and editing of the data. You must include a range of suitable data (e.g. patient information) where required, sample size of ten records minimum.

Tasks

1. Draw an entity-relationship diagram based on the information given for the surgery; clearly outline the attributes and properties
2. Develop a database and perform the following tasks:
    (a) Calculate the return date of the books that are loaned
    (b) Produce a report of all patients that are seeing a particular doctor for the day
    (c) Calculate how many patients see the nurse in any one day.
3. Edit the database to incorporate the following tasks:
    (a) Include a new doctor – Dr Mark Herring
    (b) Correct the entry for Jan Greene to read Mrs Janette Greene, and change her title to Practice Manager
    (c) Create a new table that identifies the medication and advice given to patients that have been seen by the doctor (this information should not be linked to the patient, as consultations are confidential but it can be linked to the doctor).

4.6 Activity Six – Case study: The Organic Food Company

You work as a consultant for the Organic Food Company, which sells a range of produce across East Anglia. Your role is quite diverse, as you are employed to look at a range of processes extending across sales, marketing, operations, distribution and finance.

John Peters the owner of the company has asked you to look at the 'sales ordering system', which includes the stock control and distribution systems, to see if they can all be integrated and centralized using a database system. To help with the design of the database John has provided you with the following information.

Current system – sales ordering

The company accepts sales order requests from individual household customers, shops and restaurants. Orders can be taken by phone, fax or post (currently there is no on-line provision).

When a customer telephones the following details are taken:

- Customer number
- Contact details (name, address, telephone number)
- Order details (product description, quantity, price)
- Delivery details (to be picked up or delivered and when)
- Payment options (on account, credit card, payment on receipt of order).

If the details are taken by phone, the information is recorded on a sales order sheet (template provided); if, however, the customer sends in a postal or fax request the details are taken from this source. Confirmation of the order – items in stock, prices, delivery etc. – is carried out within the hour as there is no electronic system of linking orders to stock and delivery.

Current system – stock control and delivery

When a sales order is received the information is passed onto personnel within the stock and delivery department. They check to see if the stock items are readily available on the premises and when these items can be delivered. This involves looking at the delivery schedule on the wall to see how many customers have been booked in over the course of the week and for which specific postcode areas. If the items are available the customer will be contacted and the delivery date and time confirmed.

If the items are out of stock, a supplier will be contacted to confirm delivery of the items to the company. The customer will then be contacted and informed of stock status and estimated delivery.

Supplier details:

The company is supplied by two organic distributors:

- 'Caddy's' supplies all the fresh organic dairy, meat, fruit and vegetables
- 'Howe's' supplies the processed and packaged organic food and drinks.

Deliveries from Caddy's is on a daily basis, Howe's deliver items twice a week, both suppliers are located within a 20 mile radius of the company.

Tasks

1. Produce an entity-relationship diagram for the sales ordering system to include stock control and distribution. (Entity descriptions and an attribute list should be included)
2. Carry out normalization on the sales order sheet (to 3NF)
3. Design a database to incorporate the functions of the sales ordering system.

Sales order sheet

Date: --/--/----	Customer number: ------/---

Contact details

Name:
Address:

_____

_____

_____

_____

_____

Telephone number:

Order details

Stock number: ------------/-----   Quantity ☐

Description:

Unit cost: £   Total cost: £

Delivery details

Address to be delivered to:

Invoice address (if different):

Payment options

Please tick

Cash ☐   Cheque ☐   Switch/Delta ☐   Credit card ☐   Account ☐

Paid ☐   Payment on receipt ☐

TOTAL INVOICE COST £

Figure 4.2

Example of normalizing a data set
Unnormalized data set

Student number*	Student name	Module code	Module name	Grade	Lecturer	Room number
CP123/OP	Greene	C122	IS	D	Jenkins	B33
CP938/CP	Jacobs	C123	Hardware	M	Smith	B33
CF489/LP	Browne	C124	Software	M	Osborne	B32
CP311/CP	Peters	C111	Internet	P	Chives	B32
CR399/CP	Porter	C110	Web design	D	Crouch	B33
CD478/JP	Graham	C107	Multimedia	M	Waters	B31
CR678/LP	Denver	C106	Networking	P	Rowan	B31

* Primary key

Step One – First normal form (1NF)

Student number	Student name
CP123/OP	Greene
CP938/CP	Jacobs
CF489/LP	Browne
CP311/CP	Peters
CR399/CP	Porter
CD478/JP	Graham
CR678/LP	Denver

Student number	Module code	Module name	Grade	Lecturer	Room number
CP123/OP	C122	IS	D	Jenkins	B33
CP938/CP	C123	Hardware	M	Smith	B33
CF489/LP	C124	Software	M	Osborne	B32
CP311/CP	C111	Internet	P	Chives	B32
CR399/CP	C110	Web design	D	Crouch	B33
CD478/JP	C107	Multimedia	M	Waters	B31
CR678/LP	C106	Networking	P	Rowan	B31

Step Two – Second normal form (2NF)

Student number	Student name
CP123/OP	Greene
CP938/CP	Jacobs
CF489/LP	Browne
CP311/CP	Peters
CR399/CP	Porter
CD478/JP	Graham
CR678/LP	Denver

Student number	Module code	Grade
CP123/OP	C122	D
CP938/CP	C123	M
CF489/LP	C124	M
CP311/CP	C111	P
CR399/CP	C110	D
CD478/JP	C107	M
CR678/LP	C106	P

Module code	Module name	Lecturer	Room number
C122	IS	Jenkins	B33
C123	Hardware	Smith	B33
C124	Software	Osborne	B32
C111	Internet	Chives	B32
C110	Web design	Crouch	B33
C107	Multimedia	Waters	B31
C106	Networking	Rowan	B31

Step Three – Third normal form (3NF)

Student number	Student name
CP123/OP	Greene
CP938/CP	Jacobs
CF489/LP	Browne
CP311/CP	Peters
CR399/CP	Porter
CD478/JP	Graham
CR678/LP	Denver

Student number	Module code	Grade
CP123/OP	C122	D
CP938/CP	C123	M
CF489/LP	C124	M
CP311/CP	C111	P
CR399/CP	C110	D
CD478/JP	C107	M
CR678/LP	C106	P

Module code	Module name	Lecturer
C122	IS	Jenkins
C123	Hardware	Smith
C124	Software	Osborne
C111	Internet	Chives
C110	Web design	Crouch
C107	Multimedia	Waters
C106	Networking	Rowan

Lecturer	Room number
Jenkins	B33
Smith	B33
Osborne	B32
Chives	B32
Crouch	B33
Waters	B31
Rowan	B31

4.7 Activity Seven – Database application cycle

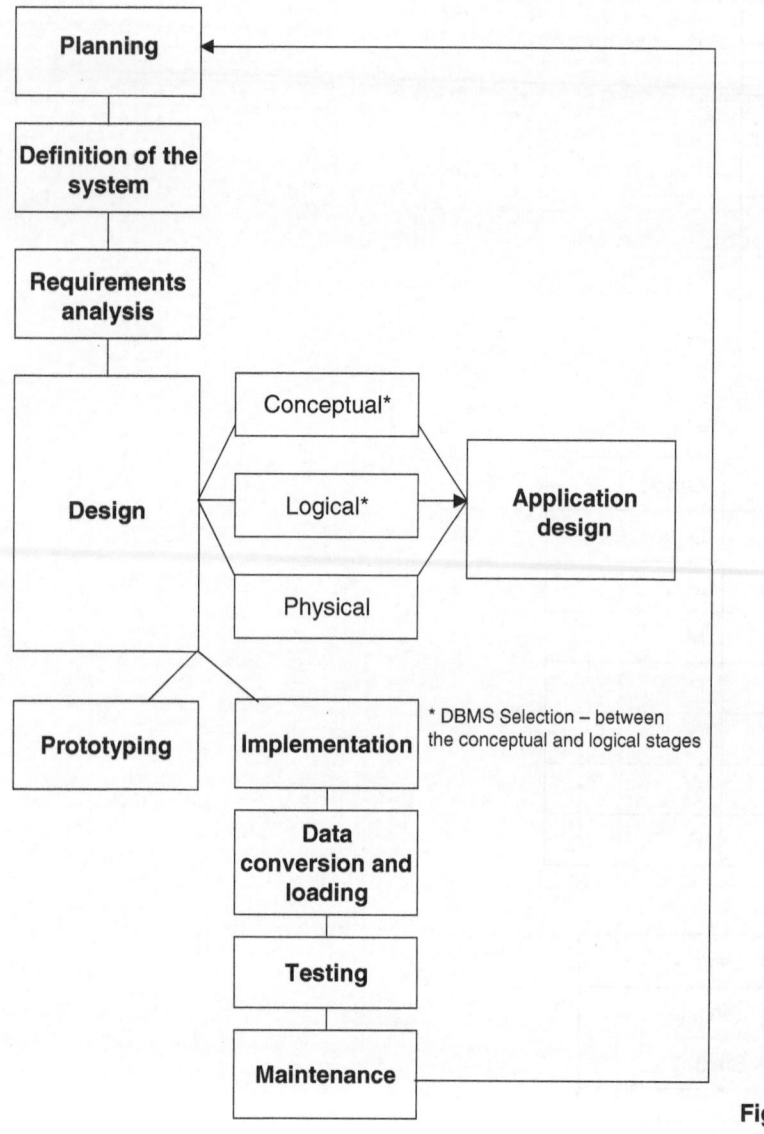

* DBMS Selection – between the conceptual and logical stages

Figure 4.3

Tasks

1. Provide a summary of each of the stages in the database application cycle.
2. Why is it important to prototype a database and what are the potential consequences if this is not carried out?
3. What testing strategies can be employed?
4. What issues may arise during the data conversion and loading stage, and what steps can be taken to address these?

4.8 Activity Eight – SQL exercises

1. Using the table and column names listed below use SQL to define a staff table with the create command.

Table name	Column name	Data type	Length	Column
Employee	Number	Decimal	6	Primary key
	Name	Variable character	15	
	Department	Character	8	
	Salary	Decimal	6	

2. Using the data below use the insert command or a form to store data in the employee table.
3. Write a selection query to list all of the employee data with department equal to IT in descending order by salary.

Data for the employee table:

Number	Name	Department	Salary
123456	Jones P	Sales	13 654
364778	Rush K	Accounts	21 789
589937	Peters L	IT	32 100
833309	Jacobs D	IT	18 699
347288	Mann P	IT	12 980
484772	Crighton S	Sales	30 544

4. Retrieve and print the employee data in order by number.

4.9 Activity Nine – Data mining

Data mining is a generic term that covers a range of technologies. The actual process of 'mining' data refers to the extraction of information through tests, analysis, rules and heuristics. Information is sorted and processed from a data set in the hope of finding new information or data anomalies that may have remained undiscovered.

Tasks

1. The article below 'Minimize data mining time to solution' discusses the benefits of using 'Clementine' as a data mining workbench. Read through the article and describe the benefits of using such a tool/workbench.

Minimize data mining time to solution

Data mining with Clementine is a business process designed to minimize the time it takes to find solutions to business problems. Clementine's powerful visual interface – combined with time-saving process support tools – enables you to use business expertise to quickly interact with your data and discover solutions in the shortest amount of time possible. Clementine supports the entire data mining process, including data access, transformations, modelling, evaluation, and solution deployment. Clementine not only supports the entire data mining process from beginning to end, it supports the industry-standard process – CRISP-DM.

Quickly discover solutions using train-of-thought analysis

The interactive, visual approach to data mining is the key to Clementine's ability to minimize time to solution. You search for a solution to your business problem by creating and interacting with a data mining stream – a visual map of the entire data mining process. This visual approach makes it easy to see every step in the process clearly – and enables you to apply business expertise to quickly explore hunches or ideas by interacting with the stream. Clementine visual data mining makes 'train-of-thought' analysis possible so you focus on solving problems rather than performing technical tasks, such as writing code. Other data mining workbenches force you to stop and write code – breaking your train of thought and dramatically slowing your progress toward a solution.

Visualization techniques accelerate solution discovery

Clementine's wide range of data visualization techniques also accelerate progress toward a solution by helping you understand key relationships in your data and guiding the way to the best results. Dramatically speed up data exploration by visually analyzing complex data sets using the data audit node. Explore multi-dimensional data with 3-D, panel, animation, and other types of data visualization. Quickly understand the results of Kohonen, K-Means, and Two-Step cluster models in the cluster viewer.

Spot characteristics and patterns at a glance with Clementine's interactive graphs. Then 'query by mouse' to explore these patterns by selecting subsets of data or deriving new variables on the fly from discoveries made within the graph.

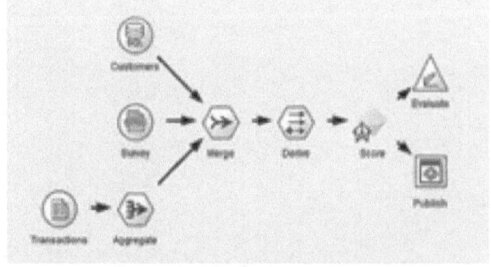

Clementine's visual approach makes it easy to see and interact with every step in the business process of data mining – from accessing data wherever it resides to deploying results to the point of decision making.

Figure 4.4 *From http://www.spss.com. Text and figure reproduced by permission of SPSS Inc.*

2. What other techniques can be used to mine data?
3. Carry out research to identify one organization that has used data mining techniques and discuss the benefits and issues of using such techniques.

5 Network assignment

5.1 What you are required to do

As an independent IT consultant, you have been asked to advise on a network connection to suit a small company. This company is outlined below. After an unhappy experience with an ISP, they have decided to run their own internet connection and servers. Rather than specify a single solution, you are asked to provide the pros and cons of different solutions and to present these at a meeting with the company, where the decision about the next step will be made. It is *not* intended at this stage to decide on the final solution, but simply to present to the company the main options before a specialist network company is employed.

Hammer and Tong Co. Ltd

Birmingham-based Hammer and Tong Co. Ltd supplies small tools to the engineering industry. They were set up in 1896 and the firm has been in the same family ever since. Their distribution is almost exclusively mail order. In 1998 they had an e-commerce package put together for them by an external company; this web-based system generally works very well and Hammer and Tong have come to rely on this form of business. Employing 30 people, they do not have a dedicated IT specialist on site but five of the staff are able to update databases and to carry out day-to-day running of a computer system. They already have a peer-to-peer LAN that connects to their old ISP using a single modem, used mainly for email and updating the e-commerce database. Currently all the network traffic is handled by their ISP but unfortunately they have recently become unreliable. They had started to blame the e-commerce package for a series of failures but Hammer and Tong believe that the fault lies with the ISP. Hammer and Tong have decided not to rely on a third party for what has become a core part of their business so they require a system to provide the following performance:

1. On average, 40 orders are processed each day, the peak being approximately 100 orders on a Monday morning reducing to about five on a Friday afternoon. Previous experience shows that for every order, potential customers refer to their website at least ten times giving an average hit rate of 400 hits per day but a peak of 1000 hits on a Monday morning.
2. The website software has been designed to load quickly so has a low bandwidth requirement, a typical 'look' at the website requiring about 80 kb of data including images. High resolution images of products are available but most customers are regulars so these are not called upon very often. These images are on average 140 kb each.
3. Hammer and Tong use a third party banking house to collect on-line payments but they are concerned that hackers could gain entry to their system so a good level of security is required.
4. Profitability in the engineering sector has been lower than previous years. While the cost of the network link is not the first priority, it will be examined very carefully before any decision is made.

5.2 What you must deliver

1. Recommendations about the network connection to the outside world
2. Recommendations on security.

5.3 Possible solutions

High bandwidth network connection possibilities:

- Cable
- ISDN
- DSL
- Satellite
- Leased line.

Cable
Speed
The cable system was designed and installed for one-way TV signal delivery, the download bandwidth is high but uploading data is often difficult, not due to the cabling bandwidth but design difficulties with the installed equipment at the cable company. There are specifications such as MCNS (Multimedia Cable Network System) and DOCSIS (Data Over Cable Service Interface Specification) but not all cable companies can offer these. Most are able to supply good bandwidth for download but require a modem for uploads. This situation will change, possibly in the near future, but the cable companies have been slow to offer high speed net access.

Security
To connect to a client, most cable companies use the 75 ohm coaxial cable that is similar to the old 'thick' Ethernet cable. Cable modems generally use 10 baseT (Ethernet) and run TCP/IP over what is in effect a small LAN. This means that one connection is shared with other users, creating a security risk so a firewall would have to be used to avoid this risk.

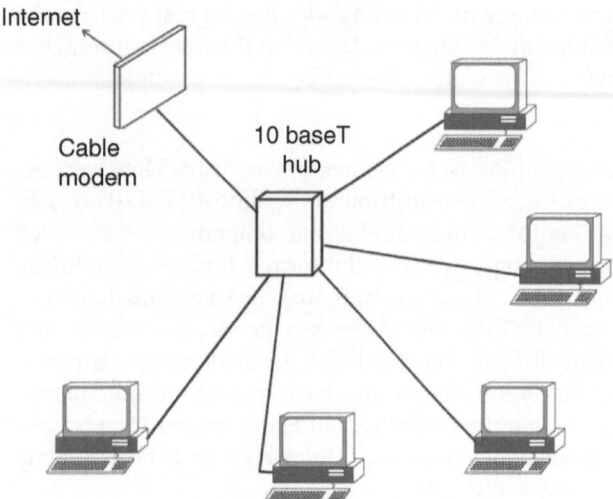

Figure 5.1 *Insecure cable connection*

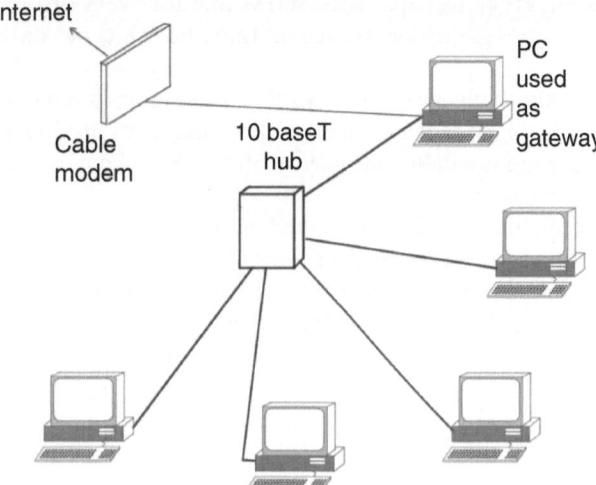

Figure 5.2 *Secure cable connection*

Effect on the LAN topology
A cable modem connection will allow several computers to be connected (often not allowed with xDSL). This would be insecure if it was done simply via a 10 baseT hub; a better solution would be to attach a PC to act as a gateway and to attach this to a hub. The gateway can act as a firewall.

Cable is aimed at heavy 'home' use and many cable providers will block the TCP/IP ports used to run web servers etc. Currently a cable connection is not a viable solution but may be in the future particularly in those areas where the cable laid into premises was fibre.

ISDN

ISDN was designed as a digital telephone system but has been adopted to run data communications. It requires that 'calls' are set up for each communication session so is not suitable for running a web service.

DSL

The POTS (Plain Old Telephone System) was designed for a bandwidth of about 3 kHz. This bandwidth applies to the whole channel, i.e. the connection from one telephone to the next. It was realized that the 'local loop', i.e. the copper wire from the exchange to the subscriber, had a far higher bandwidth provided it is short (a few kilometres depending on which DSL standard is used). DSL or xDSL is the name given to a range of communication standards designed to carry voice and data traffic over the local loop. It is not available in all areas, a check can be run at www.bt.com/adsl to find out. The current xDSL standard that is becoming available in the UK is ADSL and the best business package will provide a bandwidth of 1 Mbits/sec download and 256 kbits/sec upload but some providers will block the TCP/IP ports used to run web servers etc.

Satellite

Although download speeds can be as high as 4 Mbits/sec, the upload speed of a satellite link is usually via a normal modem so is not suitable for running a web server.

Leased line

Speed

A leased line is a permanently made, 'always on' connection; the leasing is usually from a telecommunications company. It is normal to have a line to an ISP that is in turn connected via high bandwidth leased lines to the rest of the internet backbone. The most basic leased line will provide 64 kbits/sec increasing in 64 k increments. More expensive lines running SONET (Synchronous Optical Networking) run at 51.84 Mbits/sec and increase in 51.84 Mb increments. Unlike DSL and cable modems, leased lines are aimed squarely at businesses and usually come with a range of permanent TCP/IP addresses and fixed domain names.

Security and its effect on the LAN topology

To achieve a reasonable degree of security a firewall is required. If this is placed between the router and the rest of the internal network, it would not be practical to run a web server. The better but more expensive option is to have a 'demilitarized zone' as in Figure 5.5. Here the web and email servers are outside the firewall that protects the internal LAN but are under the protection of a dedicated firewall of their own. This will result in 'public' and 'private' parts of the network. One of the higher costs associated with a DMZ is the requirement to have specialized security experts set up the entire network; it is not a trivial task.

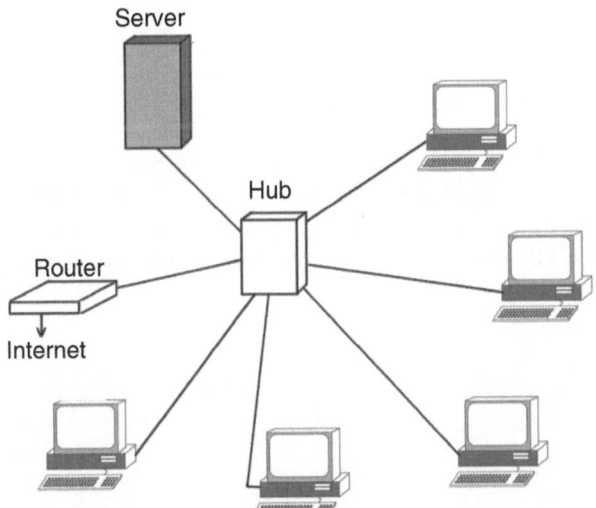

Figure 5.3 *Insecure leased line connection*

The main recommendation would be to invest in a 128 kbits/sec leased line, the additional hardware of firewalls, servers etc. and to budget for specialized people to configure the whole system. From the figures describing the likely hit rate on the website, it would seem logical to provide a minimum bandwidth of 64 kbits/sec. This is calculated from the peak hit rate of 1000 hits in a morning translating into 250 hits

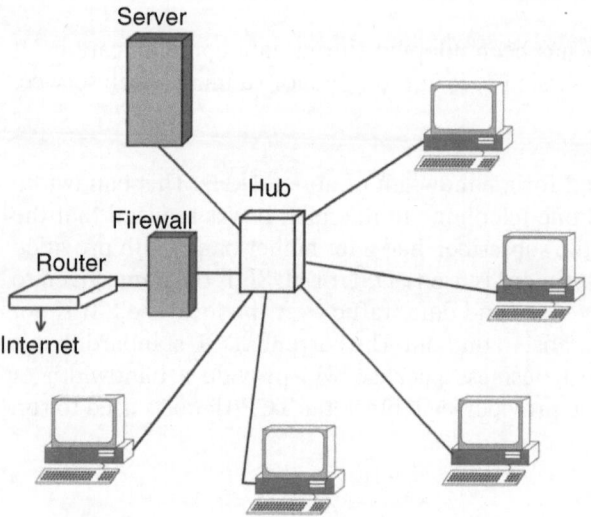

Figure 5.4 *Basic secure leased line connection*

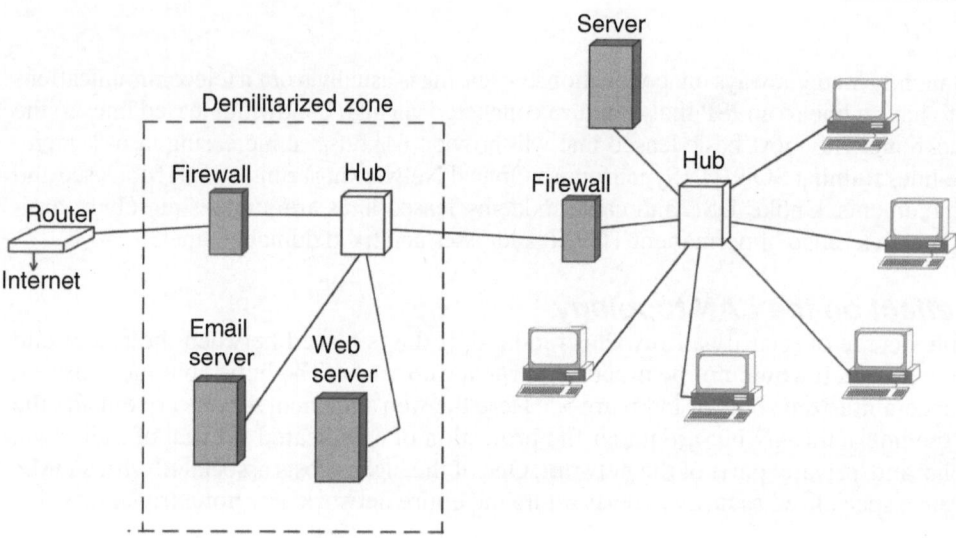

Figure 5.5 *Leased line with demilitarized zone*

per hour or about 4 hits per minute. If each hit requires the upload over the network of 80 kb of data (800 kbits) then $4 \times 800 = 3200$ kbits per minute or 53 kbits/sec is required.

This figure of 53 kbits/sec is close to the maximum bandwidth of the line and does not take into consideration any other traffic occurring at this peak time such as customers requiring the higher resolution image etc. If the e-business is to grow it is vital that potential customers very rarely find they have a slow response. For this reason it would be better to specify 128 kbits/sec as the minimum bandwidth and to take an option on increasing this to 256 kbits/sec when required.

6 Personal skills development

6.1 Activity One – Planning and reviewing

Gantt charts provide a diagrammatic overview of the sequence and timing of events for a particular task-based project. They identify and predict when activities will start and end and also examine the ordering of particular tasks.

	January			February				March		
Week no.	1	2	3	4	5	6	7	8	9	10
Activity:										
Carry out investigation			→							
Collect all the evidence together			→							
Analyse the results					→		→			
Write up results					→					
Provide some recommendations							→	→		
Present findings									→	
Gather feedback										→
Review the process										→

Figure 6.1

Tasks

The Gantt chart above provides a ten-week overview of how an activity can be planned and delivered.

1. Examine the chart and critically analyse the sequence and time allocation for each of the activities; criteria to consider are:
   • Has too much or too little time been allocated to a task?
   • Is the sequence of activities set out in a logical way?
   • Have any activities been missed out (based on your own experiences of planning an activity)?
2. Produce a revised Gantt chart based on your analysis of task 1.
3. For a given HNC/HND assessment produce a Gantt chart that illustrates the planning process you will undertake for completing the work.

6.2 Activity Two – Report writing skills

Tasks

1. Carry out research, collecting evidence for one of the following topic areas:
    (i)   The need for good transferable skills within the IT industry
    (ii)   The benefits of 'e-learning'
    (iii) Traditional class-based learning versus e-learning
    (iv)   Higher education options – vocational versus academic learning.
2. Produce a 1500 word report based on your findings.
3. Produce a short review evaluating the process of evidence collection and report writing. Consider the following points:
    • How easy or difficult it is to find the information
    • Where the information was found – what resources were used and why
    • Structuring the report.

Report format template

Section layout

1.0 Introduction: the introduction should provide a short summary of the overall focus and content of the report

2.0 Procedures: identification of any procedures used to collect, collate, analyse and present information

3.0 Main findings: the main findings section is where the bulk of the report content should be placed. The main findings section should be broken down into task, action or research areas. Each area of the findings section should put forward arguments or statements supported by research and analysis. The main findings section can be broken down further into subsections, for example:
    3.1   Marketing resources
        3.1.1   Staffing levels
        3.1.2   ICT support
        3.1.3   Staff training

4.0 Conclusions: the conclusion section should bring together all of the items discussed within the main findings section and provide a summary of the key areas identified

5.0 Recommendations: this section is solution based, providing the subjects of the report with proposals as to how they can move forward with the report objective. For example, recommendations for staff training could include:
    1. Provide residential management training to all supervisors and section managers
    2. Offer in-house ICT training programmes to all data entry clerks within the marketing department
    3. Set up staff training services on a rotary basis of three employees each week for eight weeks.

6.0 References: this section should identify and give credit for all information sources used to include books, magazines or journals, other documents or reports, and the Internet etc.

7.0 Appendices: this section will provide supporting documentation to give additionality to the report content. Appendices could include lists of facts and figures, leaflets, downloaded information, photocopied material etc.

6.3 Activity Three – SWOT analysis

A SWOT analysis:

• Strengths
• Weaknesses
• Opportunities
• Threats

is modelled on four key elements, which together provide a holistic view of a particular proposal or project. For example:

should the stock control system be upgraded in the warehouse department?

Stock control SWOT analysis:

Strengths	Weaknesses
• Would reduce the amount of paperwork • Allow automatic stock ordering • Provide automatic tracking of the stock distribution in the warehouse • Information can be accessed by all stock personnel • Would reduce overheads by 5%	• Initial financial outlay • Training of all warehousing personnel
Opportunities	**Threats**
• Stock system can be integrated into other department systems • Access to stock information by all store personnel • Links into supplier systems	• Incompatibility with other existing systems in the store

Figure 6.2

In order to assess whether or not the proposal would be viable the strengths should outweigh the weaknesses and the opportunities should exceed the threats.

Tasks
1. Provide a short summary on the use of a SWOT analysis as a tool for planning and decision making.
2. Is there anything else that needs to be considered and potentially added to any of the four sectors?
3. In your opinion, what measures could be taken if the weaknesses did outweigh the strengths and the threats exceeded the opportunities in the given example?
4. Produce a SWOT analysis based on your decision to do and complete an HNC/HND qualification.

6.4 Activity Four – Team roles

Figure 6.3 is a list of Belbin's team roles and descriptions:

Tasks
1. Look at the list of roles and identify one that is most characteristic of yourself
2. Write down three instances where you have demonstrated these characteristics
3. List two occasions when you have been aware that you have displayed one or all of the allowable weakness/es for that particular role
4. Ask three other people to select a role that they feel is characteristic of you (do not disclose your own choice)
5. Discuss the choices made and compare these to your own role selection – do their choices surprise you?

Roles and Descriptions		
	Team-Role Contribution	Allowable Weaknesses
Plant	Creative, imaginative, unorthodox. Solves difficult problems.	Ignores incidentals. Too pre-occupied with own thoughts to communicate effectively.
Resource Investigator	Extrovert, enthusiastic, communicative. Explores opportunities. Develops contacts.	Over-optimistic. Can lose interest once initial enthusiasm has passed.
Co-ordinator	Mature, confident. Clarifies goals. Brings other people together to promote team discussions.	Can be seen as manipulative. Offloads personal work.
Shaper	Challenging, dynamic, thrives on pressure. Has the drive and courage to overcome obstacles.	Prone to provocation. Liable to offend others.
Monitor Evaluator	Serious minded, strategic and discerning. Sees all options. Judges accurately.	Can lack drive and ability to inspire others.
Teamworker	Co-operative, mild, perceptive and diplomatic. Listens, builds, averts friction.	Indecisive in crunch situations.
Implementer	Disciplined, reliable, conservative in habits. A capacity for taking practical steps and actions.	Somewhat inflexible. Slow to respond to new possibilities.
Completer Finisher	Painstaking, conscientious, anxious. Searches out errors and omissions. Delivers on time.	Inclined to worry unduly. Reluctant to let others into own job.
Specialist	Single-minded, self-starting, dedicated. Provides knowledge and skills in rare supply.	Contributes on only a limited front. Dwells on specialised personal interests.

Figure 6.3 *From e-interplace SPI report, by permission of Belbin Associates UK*

6.5 Activity Five – Verbal communication

Verbal communication is one category of communication that is used to transmit information. The transmission of verbal information depends upon a range of physical and resource factors that include time, cost, the sender and recipient of the information, the transmission tool and the environment that the message is delivered in.

1. Complete the following table by identifying a situation that you have been in which involved that particular category of verbal communication:

Categories	Example of use
Negotiating	
Persuading	
Debating	
Delegating	
Challenging	
Advising	
Arguing	
Apologizing	

6.6 Activity Six – Creating a CV

A curriculum vitae is an important document as it provides you with a detailed summary of your qualifications, skills and achievements (work based, academical or social). CVs are used to complement applications for a job or entrance into higher education or training.

Curriculum Vitae

Name
Address
Contact details (telephone and email address)

Date of birth

Education and qualifications:

Further education details	Qualifications	Dates
		To: From:

Secondary school details	Qualifications	Dates
		To: From:

Other certificates and/or qualifications:

Employment/Work experience:

Employer details	Responsibilities	Dates
		To: From:

Professional memberships:

E.g. BCS (British Computer Society)

Skills and abilities:

E.g. Use of applications software, specify some packages, specify programming languages that you can use, state whether or not you have developed or designed anything, e.g. database, website etc.

Passed driving test, team captain, voluntary work etc.

Hobbies:

Try to put down a range of activities.

References:

Put down the details (name, address and contact details of two referees – usually a teacher and an employer, or somebody who knows you but is not related).

There are a number of guidelines about what a CV should contain, its layout and length but ultimately a CV should provide facts and information about your abilities. Generally a CV should be no longer than two A4 sheets, some stipulate only one, but this can be difficult if you are experienced and qualified in a number of areas.

A CV is personal and it should be unique to you; some people have different CVs promoting different skill aspects depending upon the application. One thing to remember is that your CV may end up in a pile with another hundred; what will make them read yours?

Tasks
1. Based on the template headings provided produce an up-to-date CV. Try to include information under each heading (if you have never had a job, include any work experience or placements that you have done)
2. Write down a list of all your skills and abilities, and identify what you have achieved. You could produce one academic CV and one work-based CV.

6.7 Activity Seven – Impact of verbal communication

The impact of verbal communication can be very strong, and as a result change your relationship with whom-ever you are speaking to momentarily or permanently.

1. Think of a situation where you have been the receiver of verbal communication recently and think about how you felt towards the receiver under the following situations:
   (i)   Being given instructions to carry out a task
   (ii) Being congratulated
   (iii) Being advised
   (iv) Being confided in.

Verbal communication is probably the most important and most used form of communication. As a result verbal communication has a great deal of benefits over other communication methods.

2. Identify six advantages of using verbal communication over written and visual communication formats
3. Identify any limitations of using verbal communication.

Interviews are a good example of testing how well you communicate verbally because you have to present yourself professionally and competently. Your skills of trying to persuade and convince the interviewer that you are the best candidate for the job are required from the moment you step in to the moment you walk out.

You are attending an interview for a part-time job at your local college within the IT department, sup-porting the help desk. The position requires somebody to work as part of a team to help support students and staff with queries, and assist with software installations and some configurations. The job description also mentions somebody with possible database and web-based experience.

4. Draw up a list of topics that you feel you could discuss at the interview in relation to your own experi-ences and skills
5. What further information might you require at the interview?

6.8 Activity Eight – Producing written documents

The following are samples of documents that are used in organizations, they include:

- Letter
- Memorandum
- Report.

1. Using similar templates produce each of the documents using applications software (do not use the wizard function)
2. Once the templates have been set up and saved with the appropriate headings for inserting dates, addresses, subject headings and closure sections print out a copy of each
3. Using the information given complete the template documents and save the new documents as different files.

Letter content

A formal letter needs to be sent out to:

Mr and Mrs Richard Wright
36 Honeypot Lane
Colney
Norfolk
NR12 4ER

The letter is from:

Norfolk Gardening Society
The Priory
Cheddar Way
Norfolk
NR4 9LO

Date: Monday 1 December 2003
A suitable introductory heading is required to open the letter.

Mr and Mrs Wright have won first prize in the Norfolk Gardening competition. They have won £250 of gardening vouchers and a plaque that will be presented to them in June at the Society's annual dinner dance. Invitations are enclosed, reply requested ASAP.

Mel Gladding
Chair of the Society

Memorandum content

A memo needs to be sent to Michael Peterson, Director of Finance, and copied to Paul Graham, Tim Phillips, Amanda Greene and Jenny Crop (all finance administrators). Dated 30 November 2003, the memo is regarding the team building weekend.

Content: Just to remind you all that the finance team building weekend is taking place this May Day Bank Holiday. Can we all arrange to meet outside the foyer at 10:00 where transportation will be waiting. Walking boots are essential.

Report content

You have been asked to prepare a report as part of your 'personal skills development' unit on the following topic area.

Computer gaming has changed radically over the years from an isolated activity to a social and interactive pastime.

Discuss.
    For the report you will need to research the following areas:

(i)   Over a set period of time, e.g. ten years, identify how computer gaming has changed in terms of hardware and software developments and technology
(ii) Describe recent developments in the areas of: games consoles, on-line gaming, interactive software and interactive facilities such as companies offering these facilities
(iii) Gather statistical figures about sales of games consoles
(iv) Conclusions can be based on personal opinions about the social and interactive aspects of computer gaming. Social implications of computer gaming, e.g. too much time wasted, or the future of computer gaming.

Producing written documents
Sample letter

Sender's information

Mr Spencer James
4 Toad Cottage
Armley
NR32 4DD

If there is no letterhead the address could go to the right-hand margin: ⟶

Recipient information

The Royal Aircraft Club
Highbury House
Staunton
Essex
CO31 7JD

Date
29 April 2003

Reference Number: (if applicable)

Salutation: For the attention of the Company Secretary

Introduction: Renewal of club membership

Content:

_____

_____

_____

_____

_____

_____

Closure:

Yours faithfully

Spencer James

Memorandum

To: Jean – Sales Manager

From: Carol – Marketing Manager

CC: Mark – Marketing Director

Date: 22/02/2003

Re: Launch of new marketing campaign

Body of text would be displayed here

Just a reminder that the room for the marketing campaign meeting has changed from 1b to 2d on the second floor.

Don't forget the biscuits!

Sample report

Marketing campaign
2003

IMPLEMENTATION STRATEGIES

Jean Dye

22 February 2003
Commissioned for Sales and Marketing

Contents

Introduction .... Page 1
Ideas for marketing campaign .... Page 2
Costs .... Page 4
Resources .... Page 7
Target audience .... Page 10

1.0 Introduction: the introduction should provide a short summary of the overall focus and content of the report

2.0 Procedures: identification of any procedures used to collect, collate, analyse and present information

3.0 Main findings: the main findings section is where the bulk of the report content should be placed. The main findings section should be broken down into task, action or research areas. Each area of the findings section should put forward arguments or statements supported by research and analysis. The main findings section can be broken down further into subsections, for example:

    3.1 Marketing campaign

        3.1.1 Promotional activities

        3.1.2 Dates of launch

        3.1.3 Target audiences

4.0 Conclusions: the conclusion section should bring together all of the items discussed within the main findings section and provide a summary of the key areas identified

5.0 Recommendations: this section is solution based, providing the subjects of the report with proposals as to how they can move forward with the report objective.

    For example, recommendations for promotional activities could include:

    (i) Getting prices for local radio air time slots

    (ii) Identifying competitor strategies for marketing of the particular product

    (iii) Set up launch days for the new product

6.0 References: this section should identify and give credit for all information sources used to include books, magazines or journals, other documents or reports, and the internet etc.

7.0 Appendices: this section will provide supporting documentation to give additionality to the report content. Appendices could include lists of facts and figures, leaflets, downloaded information, photocopied material etc.

6.9 Activity Nine – Using charts and graphs

Graphs and charts are used to provide visual support to data and tables providing a clear breakdown of key data components.

Using the information provided carry out the following tasks:

1. Using applications software type in the data given and produce a bar graph clearly labelling the data components
2. Produce a pie chart identifying the market share of each car
3. Produce a line graph to indicate the peaks and troughs of the market share over the period stated
4. Identify what trends or patterns have emerged from the graphs and charts produced
5. Was the information gathered from task 4 made any easier by looking at the graphs and charts or was this transparent in the table?

Market share % for car sales							
	Engine size	January	February	March	April	May	June
Car A	1.6	35	32	36	30	38	40
Car B	1.8	21	22	23	20	25	21
Car C	2.0	17	15	16	14	18	15
Car D	2.4	12	12	11	10	13	12
Car E	3.0	6	7	8	6	9	7

6.10 Activity Ten – Presentations

Good presentation skills are essential for all walks of life and especially for progression into higher education, training or employment. Most people will be expected to give a presentation either informally at school or college as part of an assessment, or formally at work in front of colleagues, at a team meeting or to management personnel.

There is always something that you can do to improve your presentation skills, these tasks are designed to look at how you behave when you are giving a presentation and hopefully can be used to improve your presentation skills.

1. In pairs assume the role of reviewers taking it in turn to talk for five minutes about a recent film that you have watched. In preparation for this write down a list of points about:
   (i)   Who the main characters were
   (ii) The plot and/or subplots
   (iii) The graphics and special effects
   (iv) The costumes
   (v)   The realism
2. As your partner is talking make comments on certain aspects of their presentation – complete a feedback sheet as shown in the example
3. Following the presentations, each person should feed back to the other
4. Deliver the review again and try to address some of the feedback given to improve and enhance the quality of your presentation.

Feedback template

Name:		Date:	
Presentation type:		Formal	Informal
Presentation title:			

Checklist	Yes:	No:
Introduction given:		
Clear and concise throughout:		
Presentation closed:		
Variation in voice tone:		
Acknowledges audience:		
Establishes eye contact:		
Good pace of presentation:		

Comments regarding delivery:
Comments regarding body language:
Comments regarding tools used (if applicable):

Areas for improvement:
1.
2.
3.
Strengths:
1.
2.
3.

6.11 Activity Eleven – Formal presentations

You have been asked to give a formal presentation based on research in one of the following topic areas:

(i)   Emerging technologies – do we really need any more?
(ii)   The Internet – friend or foe?
(iii) On-line shopping – do we really need to support high street stores?

1. You have one week to research one of the three topic areas and give a 15 minute presentation on your findings
2. Each topic researched should provide evidence in support of and against the statements given
3. The presentations should last no more than 15 minutes with time at the end for questions
4. Your presentation should try to incorporate at least one recognized tool, e.g. an OHP, video or the use of presentation software.

6.12 Activity Twelve – Freestyle presentation

It is sometimes easier to talk about a topic that you are familiar with, and do not need to research or remember facts and figures.

1. Prepare a 20 minute presentation to be delivered in front of the class. The presentation should be based on a topic that you are interested in or a hobby or pastime, e.g.:
   • A team that you support
   • A particular sporting activity
   • Your involvement with a club or society
   • A collection of memorabilia
2. For the presentation try to bring in items to support what you are doing, e.g. certificates, sports shirts, video footage, pictures, clippings, sound clips etc.

7 Quality systems

Rationale

Students will learn about the analysis and design tools used within the software development process. They will use CASE tools to produce graphical and text-based analysis and design models. The syntax and semantics of the model will be tested to ensure the quality of the product conforms to the stated requirements.
Students will:

- learn the techniques of traditional lifecycle models and object-oriented models, and how they are developed, documented and quality assured for a simulated customer specification
- use the concepts of project management to ensure the development process is planned and implemented within prescribed deadlines and checked at various milestones to ensure the verification and validation processes
- need to determine the maintenance requirements for a given software product. The need for a software maintenance contract and how 'change' can be implemented and its effects on the documentation
- need to be introduced to the basic concepts of 'quality assurance' and how it interacts across a software development project. Estimating and costing factors need to be considered along with the timing and testing issues for a software product under development
- need to develop the quality assurance tools used throughout a developing software project. They will need to hold reviews, walkthroughs and inspections and be able to produce recognized quality assurance documents
- need to be aware of the legal implications surrounding the development of a product and its associated staff. For example, the software project manager needs to approach health and safety in a strictly systematic way by assessing the risks and hazards in the workplace and taking steps to ensure that their house is in order. Also the manager needs to be aware of the issues involving storage of 'data', i.e. recorded information.

Professionalism and standards are key issues developed by any successful manager. Students need to understand the importance of these issues and how they are implemented within a quality development programme. The concept of professionalism implies taking responsibility, accounting for one's work and performing work to the highest standards. Factors like ISO, British Standards, training (CPD) etc. need to be considered in the quest for producing quality professional software development personnel.

7.1 Quality assurance

Further information

Whatever design methodology is used when developing a software system, they all need to incorporate the following main activities:

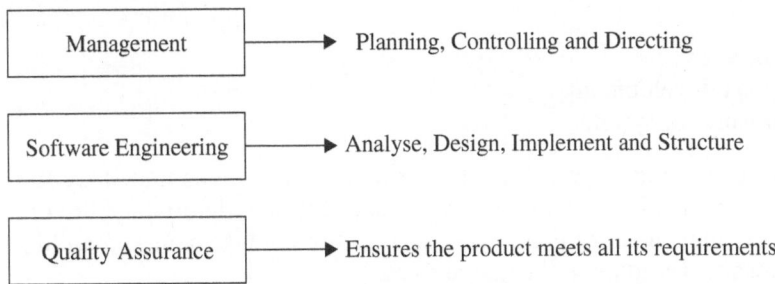

Management	→ Planning, Controlling and Directing
Software Engineering	→ Analyse, Design, Implement and Structure
Quality Assurance	→ Ensures the product meets all its requirements

Example Quality Assurance Activities

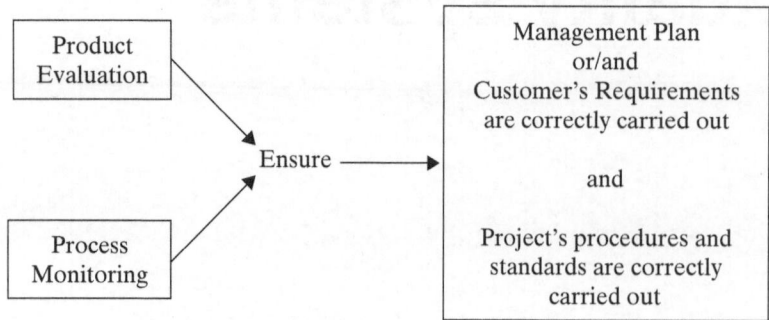

Product Evaluation ensures that standards are being followed. In an ideal situation the first products being monitored by software quality assurance should be the project's standards and procedures. The process ensures that acceptable standards exist and that compliance of these is maintained. Evaluation assures that the software product reflects the requirements of the standards as outlined in the initial plan.

Process Monitoring is a quality assurance activity that ensures the correct steps are being carried out (verification) to develop the product. This activity compares the actual steps taken with those in the original documentation to ensure they correspond.

Further information see:

http://satc.gsfc.nasa.gov/assure/agbses1.txt (also agbsec2 & 3).

Total quality management (TQM)

Total quality management is a structured system for satisfying internal and external customers and suppliers by integrating the business environment, continuous improvement, and breakthroughs with development, improvement, and maintenance cycles while changing organizational culture.

One of the keys to implementing TQM can be found in this definition. It is the idea that TQM is a structured system. In describing TQM as a structured system, I mean that it is a strategy derived from internal and external customer and supplier wants and needs that have been determined through daily management, Hoshin, Hoshin management and cross-functional management.

Pinpointing internal and external requirements allows us to continuously improve, develop, and maintain quality, cost, delivery, and morale. TQM is a system that integrates all of this activity and information.

Hoshin is a one-year plan for achieving objectives developed in conjunction with management's choice of specific targets and means in quality, cost, delivery, and morale. Accordingly, Hoshin can be described as follows:

Hoshin = Targets + Means

A target statement can be established by combining at least one direction word, performance measure, target value for the performance measure, and time period. The following table shows two elementary examples for target statements. The first example target statement reads, Decrease new product development cycle from 8 to 4 months by December 1999.

Items	Direction word	Performance measure	Target value	Time period
Example 1	Decrease	new product development cycle	from 8 to 4 months	by December 1999.
Example 2	Decrease	new product development defects	from 7 to 2%	by December 1999.

After determining a target based on this rule, the means for achieving the target should be determined. For example, to accomplish decreasing the new product development cycle from 8 to 4 months by December 1999, the means may be:

• Establish an effective development process
• Develop documentation for the product development
• Implement QFD within the quality assurance system.

Depending on the particular organization, the means will differ from other organizations that share the same targets. Customarily, there are a few means for each target. It is consequential to determine both the targets and means for Hoshin. The hierarchical management levels, such as top management and middle management, need to determine the Hoshin according to the above rules.

Some additional useful points

The guiding principle:

- Successful total quality management requires both behavioural and cultural change
- A successful TQM system brings two other management systems together with a behavioural and cultural commitment to customer quality
- Thus, TQM becomes a system within itself by default or by choice
- These three management systems must be aligned in a successful TQM initiative:
  - OM (organizational management system)
  - HRM (human resource management system) and
  - TQM (total quality management).

Two implementation approaches of TQM:

- Traditional management approach: This is the most common. A TQM is overlaid (some say forced) upon the other two systems. This approach represents the 80% failure of TQMs. In this approach TQM never becomes an accepted reality by either organizational or human resource management. It is usually seen as competition, or 'something to be tolerated'. The TQM system consumes valuable resources needed by the other systems and rejection begins to occur.
- Integrated management approach: This is the least common. A TQM is blended and balanced with existing cultural initiatives in both organizational and human resource management systems. This represents the 20% success rate of TQMs. Whether both organizational management and human resource management systems take on a 'quality management commitment' or 'join a quality management team' is not important. The principles of quality management are attended to as an important third system that blends, integrates, aligns and maximizes the other two systems to beat competition in world class quality performance. This approach can often be divided into two subchoices, depending upon managerial resources, readiness, acceptance, and competencies.

Summary

Total quality management is based on the belief that the people who are closest to the job best understand what is wrong and how to fix it. Management has the responsibility at all levels to work on the systems in which goods or services are produced.

For further information see

http://www.goalqpc.com/RESEARCH/TQMwheel.html

ISO standards update and exercise solution

An example of research into the latest ISO standards (Exercise 7.1.1).

Below are two examples from the latest standards that directly effect quality management systems.

1. Quality management principles

This document introduces the eight quality management principles on which the quality management system standards of the revised ISO 9000:2000 series are based. These principles can be used by senior management as a framework to guide their organizations towards improved performance. The principles are derived from the collective experience and knowledge of the international experts who participate in ISO Technical Committee ISO/TC 176 Quality Management and Quality Assurance, which is responsible for developing and maintaining the ISO 9000 standards.

The eight quality management principles are defined in ISO 9000:2000, *Quality management systems – Fundamentals and vocabulary*, and in ISO 9004:2000, *Quality management systems – Guidelines for performance improvements*.

Principle 1 Customer focus
Organizations depend on their customers and therefore should understand current and future customer needs, should meet customer requirements and strive to exceed customer expectations.

Principle 2 Leadership
Leaders establish unity of purpose and direction of the organization. They should create and maintain the internal environment in which people can become fully involved in achieving the organization's objectives.

Principle 3 Involvement of people
People at all levels are the essence of an organization and their full involvement enables their abilities to be used for the organization's benefit.

Principle 4 Process approach
A desired result is achieved more efficiently when activities and related resources are managed as a process.

Principle 5 System approach to management
Identifying, understanding and managing interrelated processes as a system contributes to the organization's effectiveness and efficiency in achieving its objectives.

Principle 6 Continual improvement
Continual improvement of the organization's overall performance should be a permanent objective of the organization.

Principle 7 Factual approach to decision making
Effective decisions are based on the analysis of data and information.

Principle 8 Mutually beneficial supplier relationships
An organization and its suppliers are interdependent and a mutually beneficial relationship enhances the ability of both to create value.

2. Implementing your ISO 9001:2000 quality management system

1. Identify the goals you want to achieve
   Typical goals may be:
   - Be more efficient and profitable
   - Produce products and services that consistently meet customer requirements
   - Achieve customer satisfaction
   - Increase market share
   - Maintain market share
   - Improve communications and morale in the organization
   - Reduce costs and liabilities
   - Increase confidence in the production system.
2. Identify what others expect of you
   These are the expectations of interested parties (stakeholders) such as:
   - Customers and end users
   - Employees
   - Suppliers
   - Shareholders
   - Society.
3. Obtain information about the ISO 9000 family
   - For general information, look to this brochure
   - For more detailed information, see ISO 9000:2000, ISO 9001:2000 and ISO 9004:2000
   - For supporting information, refer to the ISO website
   - For implementation case studies and news of ISO 9000 developments worldwide, read the ISO publication ISO 9000 + ISO 14000 News.
4. Apply the ISO 9000 family of standards in your management system
   Decide if you are seeking certification that your quality management system is in conformance with ISO 9001:2000 or if you are preparing to apply for a national quality award:
   - Use ISO 9001:2000 as the basis for certification
   - Use ISO 9004:2000 in conjunction with your national quality award criteria to prepare for a national quality award.
5. Obtain guidance on specific topics within the quality management system
   These topic-specific standards are:
   - ISO 10006 for project management
   - ISO 10007 for configuration management
   - ISO 10012 for measurement systems
   - ISO 10013 for quality documentation
   - ISO/TR 10014 for managing the economics of quality
   - ISO 10015 for training
   - ISO/TS 16949 for automotive suppliers
   - ISO 19011 for auditing.
6. Establish your current status, determine the gaps between your quality management system and the requirements of ISO 9001:2000

You may use one or more of the following:
- Self assessment
- Assessment by an external organization.

7. Determine the processes that are needed to supply products to your customers
   Review the requirements of the ISO 9001:2000 section on product realization to determine how they apply or do not apply to your quality management system including:
   - Customer-related processes
   - Design and/or development
   - Purchasing
   - Production and service operations
   - Control of measuring and monitoring devices.

8. Develop a plan to close the gaps in step 6 and to develop the processes in step 7
   Identify actions needed to close the gaps, allocate resources to perform these actions, assign responsibilities and establish a schedule to complete the needed actions. ISO 9001:2000 paragraphs 4.1 and 7.1 provide the information you will need to consider when developing the plan.

9. Carry out your plan
   Proceed to implement the identified actions and track progress to your schedule.

10. Undergo periodic internal assessment
    Use ISO 19011 for guidance in auditing, auditor qualification and managing audit programmes.

11. Do you need to demonstrate conformance?
    If yes, go to step 12
    If no, go to step 13
    You may need or wish to show conformance (certification/registration) for various purposes, for example:
    - Contractual requirements
    - Market reasons or customer preference
    - Regulatory requirements
    - Risk management
    - To set a clear goal for your internal quality development (motivation).

12. Undergo independent audit
    Engage an accredited registration/certification body to perform an audit and certify that your quality management system complies with the requirements of ISO 9001:2000.

13. Continue to improve your business
    Review the effectiveness and suitability of your quality management system. ISO 9004:2000 provides a methodology for improvement.

Software failure (suggested outline example solution to Exercise 7.1.2)

Some suggested examples:

Three Mile Island

This near disaster, which occurred in 1979, nearly caused a meltdown in the nuclear reactor. When the fault occurred it caused nearly 100 alarms to go off at the same time virtually overloading the system. There was no clear prioritization of the alarms by the computer system. The fault was caused by a pressure relief valve being stuck open allowing coolant to escape and the core to overheat. The emergency system did cut in and provide extra coolant, but in the confusion the operators misunderstood the information displayed in the control room and shut off the emergency system. So without adequate cooling the core started to melt. By luck the emergency system was turned back on and a catastrophe was avoided. During the inquiry that followed much attention was focused on the poor user interface of the control panels. Below are two points from the inquiry:

- A light on the control panel indicated that the pressure relief valve was still closed when in fact it was open
- The light signalling that the emergency system was off was obscured by a maintenance tag.

Patriot missile

Failures of Patriot missiles to intercept scud missiles have been attributed to an accumulation of inaccuracies in the internal timekeeping of a computer. The system was meant to be turned off and restarted often enough for the accumulated error never to become dangerous. The smallest possible perturbation (disturbance) to the state of a digital computer (changing a bit from 0 to 1, for instance) may produce a radical response.

Atlas rocket

A single incorrect character in the specification of a control program for an Atlas rocket, carrying the first US interplanetary spacecraft, Mariner 1, ultimately caused the vehicle to veer off course. Both rocket and spacecraft had to be destroyed shortly after launch. In addition to unintentional design bugs, flaws deliberately introduced to compromise a system can cause unacceptable system behaviour.

Airbus

A Lufthansa Airbus A320 crashed after landing at Warsaw Airport, killing two and injuring 54. Safeguards in the user interface delayed the pilot's attempt to activate the spoilers and reverse thrust to avoid overrunning the runway.

Tornado

Two Royal Air Force Tornado fighter planes collided in the UK in August 1988. Four crew were killed. Both planes used autopilots that used pre-programmed cassettes; they were given identical tapes, resulting in both planes reaching the same point at the same instant.

See also:

Software Failure: avoiding the avoidable and living with the rest, by Leslie Hatton, published by Addison-Wesley, January 1998.

Further example of software failure – a case study

The London Ambulance Service Computer Aided Dispatch Service

During 1992 a new computerized dispatch system was deployed for the London Ambulance Service. The system was responsible for receiving emergency calls and dispatching ambulances based on an understanding of the call and available resources. The system included an automatic vehicle locating system and mobile data terminals which supported communications with the ambulances. It was designed to supplement the existing manual system.

As the system became operational the call traffic increased but not to unexceptional levels. The automatic vehicle locating system was soon not able to keep track of the position and status of the ambulances which led to incorrect data being stored so that:

- Units were being dispatched to calls that were not optimal
- Multiple units were being assigned to the same calls.

As a consequence of this there were a large number of exception messages which caused the system to slow down as the queues grew in size. The system responded to this by sending repeated messages and some of the information scrolled of the top of the screen and was lost from view. Ambulance crews had difficulty in using the mobile data terminals, which resulted in some being slow to respond and others not using the system at all.

Towards the end of 1992 the system finally became completely unstable. Some of the resulting responses are outlined below:

- One ambulance arrived to find the patient dead and taken away by undertakers
- One ambulance arrived 11 hours late, 5 hours after the patient had been taken to hospital by other means
- A man died as his wife made repeated calls for an ambulance to arrive.

The original design specified that a fully operational backup server should be installed, but this was not operational as it had not been fully tested. The software was incomplete and not fully tested and the implementation approach was 'high risk'. Incorrect assumptions were made during the specification process and a lack of interaction between end users and clients had knock-on consequences. More problems are outlined below:

- Lack of robustness
- Poor performance
- Poorly designed interface
- Basic errors and bugs not found or fixed.

The reason for completing this case study is to 'learn from failure', i.e. learn from previous mistakes and ensure that the problems outlined do not occur again. Also to fully understand the need for a 'quality system' and what it should incorporate in the analyse, design and implementation phases of development.

What factors contributed to such a situation?

- The resulting inquiry reported that neither the CAD system itself nor its users were ready for full implementation on 26 October.
- The CAD software was not complete, not properly tuned, and not fully tested.
- The resilience of the hardware under a full load had not been tested.
- There existed outstanding problems associated with the data communications of the system, specifically communication to and from the mobile data terminals.
- There existed some scepticism over the accuracy of the Automatic Vehicle Location System (AVLS).
- Staff, both within the Central Ambulance Control (CAC) and ambulance crews, had no confidence in the system and were not all fully trained.
- The physical changes to the layout of the control room on 26 October meant that all CAC staff were working in unfamiliar positions, without paper backup, and were less able to work with colleagues with whom they had jointly solved problems before.
- There had been no attempt to foresee fully the effect of inaccurate or incomplete data available to the system (i.e. late status reporting or vehicle locations).
- These imperfections led to an increase in the number of exception messages that would have to be dealt with and which in turn would lead to more call backs and enquiries.

The decision on 26 October to use only the computer generated resource allocations was a *high risk* move.

Further points to note

The system was developed when the National Health Service was undergoing considerable change, i.e. towards decentralization and directly financial accountable management. A lack of prior investment, drawing money away from London and resource pressures tied up with the political strategy of the day were all affecting the way the National Health Service was operating.

Case study

You are now to carry out the case study below:

The above information only gives a brief outline of the problems. You are to produce a more comprehensive report on what went wrong and what *quality issues* could be used to prevent the problems occurring again.

In order to complete this case study you need to look at the full report, which can be found on:

http://people.cs.uct.ac.za/~gaz/teach/hons/papers/lascase0.9.pdf

see also

http://www.cas.mcmaster.ca/~baber/Courses/3J03/StudentPresentations/LondonAmbulanceHyland.pdf

Its ISBN number is: 0-905133-70-6.

The report was produced by the Communications Directorate of South West Thames Regional Health Authority.

Some suggested areas that your report should contain:

- Overview of the case study
- Main problem areas found
- Problems resulting from the system faults
- Quality issues that would prevent the problems occurring
- Highlight a major area of the system specified in the report and specify how the problems could be avoided using quality analyse and design techniques
- Specify the problems outlined with 'brought-in' components
- Add a suitable conclusion.

The aim of this case study is to further develop the concepts of quality systems and how we can learn from the errors defined in the report. The use of sound software engineering practices, quality assurance and project management need to be considered during the work on the case study.

For the summary you need to study the report and look for issues that keep repeating themselves. These will form major areas of concern that should have been detected originally by good practices and why thorough testing was not always evident to locate them.

The case study is looking for a lot of your own ideas, not just copied chunks of work (plagiarized) from the original report.

Legal requirements

The Data Protection Acts (further information)

History of Data Protection Acts 1984 and 1998

- The Data Protection Act 1984 grew out of public concern about personal privacy in the face of rapidly developing computer technology.
- The Data Protection Act 1998 (the new Act) was passed in order to implement a European Data Protection Directive. This Directive set a standard for data protection throughout all the countries in the European Union and applies to personal data held in a structured way in any medium (paper, computer, microfiche, tape etc.). The new Act extends the rights of individuals and increases the responsibilities of data controllers.

Data controllers, i.e. the person or persons who determine the purposes and manner in which personal data is processed, need to be aware of the following components inherent within the Data Protection Act:

1. *Data* – information which is being processed by a computerized system, or which is recorded with the intention that it will be processed in this way. In addition, any information kept as part of a 'relevant filing system' (i.e. structured either by reference to individuals or by reference to criteria relating to individuals so that the data is readily accessible), or which is recorded with the intention that it will be processed in this way, is also data. Data can be written information, photographs, or information such as fingerprints or voice recordings.
2. *Personal data* – information which relates to a living individual who can be identified from those data, or from those data and other information which is in the possession of or is likely to come into the possession of, the data controller and includes any expression of opinion about the individual and any indication of the intentions of the data controller or any other person in respect of the individual. So, for example, application forms marked only with a number will not identify an individual and will not be deemed personal data, but put together with the list of numbers and names will do so, and so will be deemed personal data.
3. *Processing* – anything which can be done with data, i.e. obtaining, recording, holding, organizing, adapting, altering, retrieving, consulting, disclosing, aligning, combining, blocking, erasing, destroying and so on.
4. *Data subject* – an individual who is the subject of personal data. For example, within a college this will include: staff, students, suppliers of goods and services, business associates, prospective students, delegates who attended a conference/course run by your school/division/centre etc., survey respondents etc.
5. *Data processor* – any person (other than an employee of the data controller) who processes the data on behalf of the data controller.
6. *Recipient* – any person to whom the data is disclosed who is an employee or agent of the data controller, a data processor or an employee or agent of the data processor.
7. *Third party* – any person other than the data subject, the data controller, the data processor or other person authorized to process data for the data controller.

What are the main changes between the 1984 and 1998 Acts?

1. Wider definition of processing – anything which can be done with personal data is now caught by the new Act. Includes obtaining, organizing, adapting, altering, amending, storing, browsing and deleting. Processing by reference to surname, student number, postcode, county etc. is also included within the new Act. Restrictions imposed on processing personal data and processing sensitive personal data.
2. The new Act applies to personal data held on computers and personal information held in 'relevant filing systems', i.e. information held in a structured way. This can mean any medium such as paper, database, videotape, microfiche etc.
3. There are new 'fair processing' principles.
4. Personal data may be transferred to non-EEA nations only if certain conditions are met.
5. The use of automated decision-making processes on their own to make significant decisions with regard to individuals is subjected to restrictions.
6. Directors, managers, secretaries or similar officers of a body corporate, including higher education institutions, can be held liable for offences committed by their institutions.
7. Individuals can go directly to court if they believe that their rights under the Act have been breached.
8. The rights of staff and students have been enhanced.
9. Some types of processing are specifically excluded from the regulations, e.g. for journalism or artistic purposes.
10. New rules for enforcement have been introduced.
11. The data protection principles have been extended from seven principles under the 1984 Act to eight principles under the new Act.

12. Statements of intention as well as opinions about an individual will be classed as personal data.
13. Wider grounds for claims for compensation from individuals have been introduced.
14. A simplified 'notification' system will replace the registration system.
15. The Data Protection Registrar will be known as the Data Protection Commissioner.

Background to the 1998 Act

The Act came into force at the beginning of March 2000. Key changes under the Act include:

- Extending the provision to include manual records
- A new definition of sensitive personal data
- An individual right to prevent processing likely to cause damage or distress
- An individual right to prevent processing for the purpose of direct marketing
- New exemptions from notification and registration
- A direct requirement on data controllers to comply with the data protection principles whether they are required to notify under the Data Protection Act or not
- The Data Protection Registrar will now be called the Data Protection Commissioner and has powers of enforcement, a new duty to promote good practice and a power to issue codes.

Rights of the individual

Individuals have extensive rights in relation to personal data held about them. In particular, they may have the right to:

- Be notified if, what, why and by whom personal data is being processed
- Prevent processing likely to cause damage and distress
- Prevent processing for the purposes of direct marketing
- Ensure that no decision which affects them significantly is being automated
- Compensation where damage or distress is suffered; have inaccurate or damaging data corrected or destroyed
- Ask the Commissioner to make an assessment on compliance.

Right of access to information

The 1998 Act significantly extends the range of information held by a data controller on a data subject which must be supplied at the request of the data subject. If personal data is being processed, the data subject must, on request, be given a description of the nature of the data held, the purposes for which it is to be processed and the persons, or categories of person, to whom it may be disclosed.

The data controller must also give information as to the source of this data. In addition, where decisions affecting the subject are likely to be taken solely on the basis of automated processing, the data subject must also be supplied with information regarding the 'logic involved in that decision taking' provided that the data controller will not need to disclose any information which constitutes a trade secret.

The data controller is only obliged to respond to requests which are made in writing and which enclose any fee required by the controller. The data controller will then have 40 days to respond satisfactorily to the request.

Exemptions

The legislation recognizes that in certain circumstances the strict application of these rules is inappropriate – because the interests of the state and/or the public outweigh the threat to individual privacy, or because compliance with the law would be burdensome for data controllers and of little benefit to data subjects. Data processed for domestic purposes is completely exempt. Further examples are the suspension of subject access rights to confidential references given by the data controller, and (often) personal data which also identifies another individual. Restrictions on disclosure do not generally apply to, for example, disclosures for the purposes of preventing crime or where disclosure is required by a court order. Journalists have the benefit of a wide exemption where it is reasonably believed that publication of the personal data is in the public interest and that compliance with data protection rules would be incompatible with journalistic purposes. Exemptions are, however, often limited in scope and, broadly speaking, the obligation to comply with the rules should be assumed.

Some additional links:

http://www.dataprotection.gov.uk/transbord.htm
http://www.hmso.gov.uk/acts/acts1998/19980026.htm

Exercise

Get the students to review how your college/university handles the Data Protection Act, for example what are their rights, what information is available to them, how can they access this information etc.

Copyright, Designs and Patents Act 1988

Suggested solution to Exercise 5.3.1

This Act made it clear that unauthorized copying of software is illegal. Copyright law protects software authors and publishers, just as patent law protects inventors. Copyright law makes no distinction between duplicating software for sale or for free distribution. Penalties include *unlimited* fines and jail terms of up to two years.

Within an organization the following is an example of software theft:

- Copying company owned software onto privately owned computers
- Copying privately owned software onto company computers
- Copying a single copy of software onto multiple machines without buying additional licences
- Putting a single copy of software onto a LAN server without paying for additional licences.

Example infringements under the Act are:

1. Software pirate gets maximum sentence – July 1999
   As a direct result of their investigations, FAST and Kingston Trading Standards ensured a custodial sentence for a software fraudster for nine counts of copyright and trademark infringements. He had offered computer software at very low prices in a leading trade magazine under a bogus company name. When they raided his home address they discovered two CD writers and 500 pirated CDs. When he appeared in court in July 1999, the magistrates decided he should be remanded to the Crown Court for sentencing. He was eventually convicted and sentenced to two years' imprisonment.
2. Software piracy, police bust AOL UK – November 1998
   Red-faced executives at AOL's London offices were busted for using illegal software after the BSA moved following an anonymous tip-off. The UK arm of AOL was caught red-handed using unlicensed copies of Adobe Photoshop and Illustrator software at the company's office in London. The BSA has been quick to point out that AOL's misdemeanour was not premeditated. Instead, AOL was the victim of its own bad systems management, which allowed their software usage to increase without securing enough licences.
   The company held up its hand to the infringement and agreed to pay an undisclosed settlement to the BSA.

For additional information see:

http://www.hmso.gov.uk/acts/acts1988/Ukpga_19880048_en_1.htm
http://www.patent.gov.uk/copy/tribunal/triabissued.htm

Professional standards – further information

The importance of Continued Professional Development (CPD)

In a modern working environment, especially within the computing and IT industry, all employees need to be aware of the need to update their technical and managerial skills. This updating can be gained in many ways and many employers are turning to a formal system of Continued Professional Development (CPD). One of the main characteristics of CPD is that of recording the additional skills that have been obtained, but this process is often off-putting as it places an additional time constraint on the employee. In an article on the growing importance of CPD by Tome Kennie, he specifies six main reasons why CPD is vital for professional success:

- Competence: knowledge gained at the start of one's career has a short life span. If you want to progress along your career you have to continually update that knowledge and add new skills
- Consumerism: consumers are simply far more sophisticated and demanding than ever before. New skills have got to be acquired if these demands are to be met
- Litigation: claims of negligence are at an all time high, and one way to cut back is to keep professionals up to date with current codes of practice
- Standards: standards of competence are safeguarded by the professional bodies, and CPD is one way of making sure that members understand and comply with such standards

- Quality management system: there is now a great emphasis on QMS and continuous improvement. CPD ensures the training and education vital to these processes
- Competitiveness: modern businesses are now even more competitive, and there is a focus on service quality and technological innovation. Again there is a need for the development of people.

CPD is a requisite for membership of an associated professional institution and registration with the Engineering Council. The Engineering Council states that registration with it, and membership of a professional institution, obligates members to maintain and develop their professional competence. Its three point code of practice maintains that members must:

1. Demonstrate commitment to maintaining professional competence through self-managed CPD. This includes being aware of and understanding your requirements for evidence, what kinds of evidence are available and appropriate, and the need for the availability of said evidence
2. Take responsibility for and manage CPD. This includes knowing your development needs, their benchmarks, how to carry out your development action plan, recording achievements and evaluating them against needs
3. Support the learning and development of others. This includes providing support and sharing expertise with others, staying involved with the activities of your institution and making sure that your employer is also aware of the importance of CPD.

See also:

www.engc.org.uk

Additional information from the BCS

The British Computer Society is a valuable source of information for the computing professional. For example, through its BCS Review 2001 there is company and corporate information ranging from technological issues to legal requirements. The Disability Discrimination Act (DDA) is an example outlined currently, with information on what it is, Disability Rights Commission (DRC) and why it is important to you.

There is also a large section on training that includes the Industry Structured Model (ISM). The ISM is a definitive set of performance, training and development standards for the information systems industry. This is aimed at companies and helps to identify staffing requirements, create job descriptions, access competence of IS staff, plan training and more. It has measurable benefits and its use can result in:

- Increased effectiveness
- Improved quality levels
- Better focused, more cost-effective training
- Significant saving in both time and money.

The ISM is recognized worldwide as the most authoritative reference and forms the basis of the European Informatics Skills Structure (EISS). The latest release, ISM3.2, was developed and updated after several months of consultation with professionals from all areas of IS.

Under the section for Continued Professional Development (CPD) the BCS outlines a measurement strategy for people updating skills. This strategy specifies a points system for updating skills, an example is outlined below:

- Courses of study leading to a successful examination, for example Microsoft Certification qualification. 1 CPD unit per hour of study
- Interactive training where both knowledge and skills are acquired and demonstrated. ½ CPD unit per hour of study
- Structured self-study that does not contain a formal examination. ¾ CPD unit per hour of study
- Preparation of a professional paper or article. 5 CPD units
- Production and (first) delivery of a professional presentation or lecture where this is *not* part of normal work duties. 5 CPD units per hour of presentation
- Attendance only CPD, i.e. lectures, seminars, workshops etc. ½ CPD units per hour
- Other useful CPD activities (other than normal work duties). ½ CPD units per hour.

As a guide, the BCS recommends that members should seek to achieve an average of at least 20 CPD units per year.

See also:

www.bcs.org.uk

Outline suggestions to Exercise 7.1.3

1. *BCS – Student member*

    To be eligible for Student membership you must be studying a course at HNC level or higher. Once you have successfully completed a course at this level you will be eligible for Graduate membership of the Society.

    Student fees: **£49***

    * Student membership fees: There is an abated rate of *£14 for full time students in colleges and universities.* (*Note:* To qualify for Student membership you must be studying a course at HNC level or higher.) There is also a one-time fee covering the full duration of your course (max. four years) £26. Correct at 1 August 2001.

    For professional development see previous information on the BCS.

2. *Alternative institute – Institute of Analysts and Programmers*

    This Institution is the premier body dealing specifically with the needs of the working analyst or programmer. Membership is drawn from all sectors of the industry and from all sizes of business. Membership grades are:
    *   Student and Graduate membership (GradIAP)
    *   Associate membership (AMIAP)
    *   Member (MIAP)
    *   Fellow (FIAP).

The Institution's main purpose is to promote vocational skills and professionalism. Thus, in the past examinations have played a secondary role in the admission criteria. The view has been that theoretical knowledge is principally the preserve of the educational establishment, and that the Institution should concern itself more with what a programmer can do rather than what he is supposed to know. While the Institution does not wish to get 'bogged down' in a complex system of examinations, it is proposing to become more active in sponsoring courses which will teach the fundamentals of programming and analysis, and provide a sound foundation on which the novice can begin to build a professional career. There has been increasing demand for such courses since the Institution began to phase in more rigorous methods for assessing the competence of potential new recruits. While the decision to seek a Royal Charter may have been the initial impetus behind these changes, they are in the interests of everyone. As a professional body, the Institute of Analysts and Programmers will stand or fall on the professional quality of its members.

See also:

www.iap.org.uk

7.2 Quality control

Software maintenance

Further information

When any physical component goes into use, for example a car, a washing machine, a vacuum cleaner or a personal computer, one would expect that very little change would be required. The change is only intermittent and is usually to replace parts that are worn out or need modifying. This is not the same with software systems; from the moment they are validated they tend to change. Policies need to be decided whereby the decisions to effect these changes can be made. Often these are commercial decisions that are designed to emit new versions of software systems that:

*   Are less wrong
*   Require improved features and functionally for end users.

Change is potentially a major source of loss of original quality during the development of software systems. Numerous systems have undergone many changes and become so patched up that it is a wonder they serve any useful purpose.

Software evolution

Evolutions of software are things like new features, or modifications to existing ones that come about for a number of reasons. The customer may require new additions to be added to a working system as his/her business evolves due to natural expansion. These could include greater use of IT within the organization, the system needs to be more economically viable, the system needs to be more user friendly as more staff are to be involved with its use etc.

Lehman's Laws of Software Evolution

1. *The Law of Continuing Change.* Programs concerning real-world requirements either undergo continual change or become progressively less useful. This continues until it is judged more economic to replace the system than to change it.
2. *The Law of Increasing Complexity.* Changes to software systems increase the complexity of their construction and, commensurately, reduce their degree of good structure with detriment in quality unless explicit work is done to maintain the degree of good structure or improve it and thereby preserve or reduce the level of complexity and thereby enhance quality.
3. *The Law of Program Evolution.* Program evolution is subject to a dynamic process that makes it, and measures of its development and system properties, self-regulating with statistically determinable treads and invariance.
4. *The Law of Invariant Work Rate.* During the active life of a program, the overall activity rate in associated programming tasks is statistically invariant.
5. *The Law of Perceived Complexity.* During the active life of a program, the change content (changes, additions and deletions) of successive releases of the evolving program is statistically invariant.

It goes without saying that changes to 'safety-critical' software should undergo severe testing and debugging routines to ensure that the modified product still meets its original high specification requirements.

For further information see:

A Study in Software Maintenance, by Dart, Christie and Brown.

PDF and PostScript files can be found at:

http://www.sei.cmu.edu/publications/documents/93.reports/93.tr.008.html

Version control

One of the main considerations of maintaining a configuration management system in practice is the documentation process. We have stated highly in this chapter that all development aspects should be documented, but control on this documentation is essential. Outlined below are two possible (perhaps exaggerated) examples of potential problems:

1. There may be a free-for-all by developers to the documents that are being produced during the development of a system. Such an unrestricted access to write to the deliverable documentation files can lead to a divergence between system requirements, source code or data elements. In the worst case major changes to one document may occur without a collateral updating of another document which could give rise to problems with the source and objects codes and the documentation not corresponding to either of them.
2. There may be highly restrictive access to the development documentation. This could involve security checks or software/hardware password keys being required which add to the problems of accessing the documentation. Such restricted access can also be placed on attempts to modify a master copy of the source code.

The second method stated above has both technical and managerial advantages, as full control is placed on both the documentation and the source code. In some cases the organization may select a nomination of authorized staff, along which security measures to complete the system of restricted access. These fairly extreme conditions occur in a surprising number of instances with managers unaware of the possible consequences.

If a company is not using an electronic version control system then an alternative system needs to be put into place. An example client/product table is outlined below:

Client	Product	Version	Document file name (Key)
Bernese Dog Food Ltd	BERDOG	2.1	BERDOG.2.1
Bailey Brewery Ltd	BAIBRE	1.7	BAIBRE.1.7
Morgan Freight Assoc.	MORFRE1	3.4	MORFRE1.3.4
	MORFRE2	1.2	MORFRE2.1.2
Tia and Susie Dance Studio	TIASUS	1.4	TIASUS.1.4

Whatever method is employed a successful version control system is essential for the success of the configuration management process.

Suggested solution to Exercise 7.2.1 – version control

This has been completed using the ComponentSoftware RCS (Revision Control System) which monitors changes made in files that are accessed by stand-alone or networked workstations.

This system can be downloaded free for single workstations or purchased under licence for network use.

It can be found at:

http://www.componentsoftware.com/

A simple example of document control using CS-RCS revision control system:

Step 1 *Getting started*
Save a simple text file with some simple lines of text in it. Call it MyTest1.txt

Step 2 *Add a file to the Repository*
Select the file in Windows and right click the mouse button.

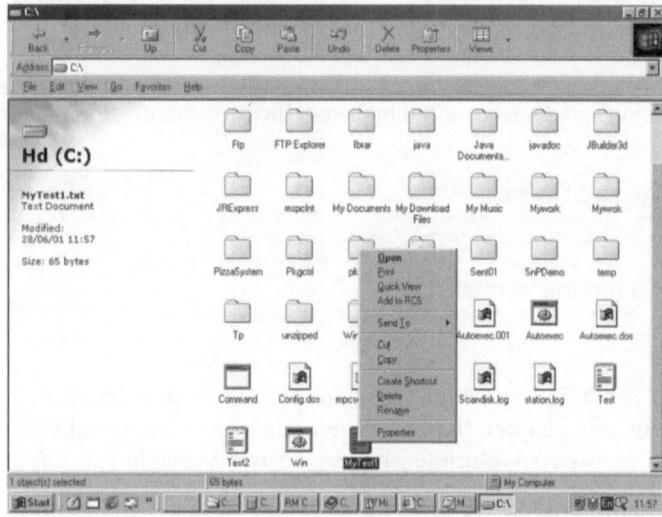

Figure 7.1

The dialog display contains RCS components required to handle the control of the document. From this select the Add to RCS.

A message box appears asking to confirm this. Click yes. Then click OK. The file has now been added to the Repository.

Step 3 *Checking the file*
To check whether MyTest1.txt changed since it was last added to the Repository:

1. From Windows Explorer, right click MyTest1.txt. The right click menu appears:

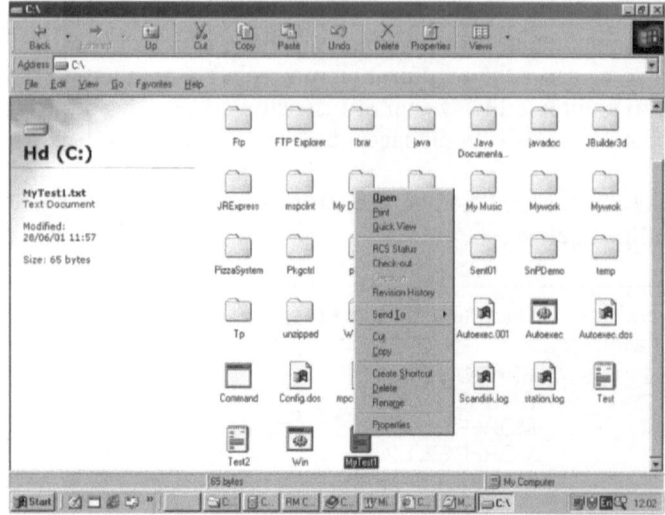

Figure 7.2

Note: As long as MyTest.txt is unchanged, the Check-in menu entry is disabled. If you installed ComponentSoftware as Personal, the Check-out menu entry is also disabled.

Next click on the RCS Status and the following message appears:

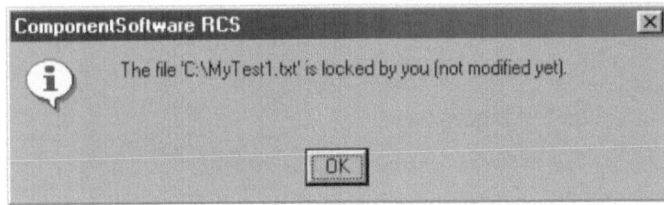

Figure 7.3

Select OK.

Step 4 *Editing the file*
Edit the file MyTest1.txt by adding some more lines. Then save the file. Then check and view the file changes in the repository.

  Right click the file in Windows Explorer
  Select RCS Status
  The following message appears:

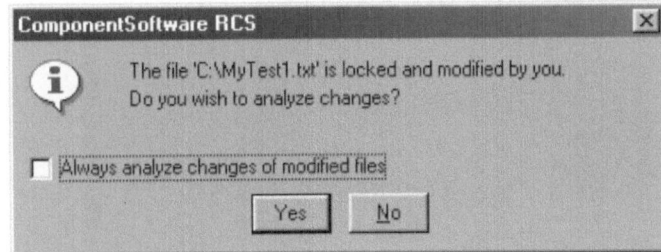

Figure 7.4

  Click Yes.

The file is displayed with the modifications and deletions noted.

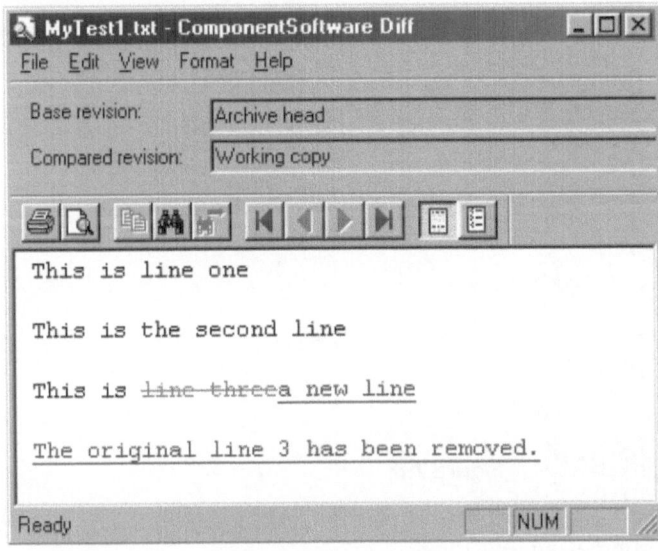

Figure 7.5

Step 5 *Checking out a file*
By checking out a file, you can undo all the changes you made to a file since it was last added to the Repository. When you check out a file, you are retrieving the latest revision of a file.

  Right click on the file MyTest1.txt
  Select Check-out

The following screen is displayed:

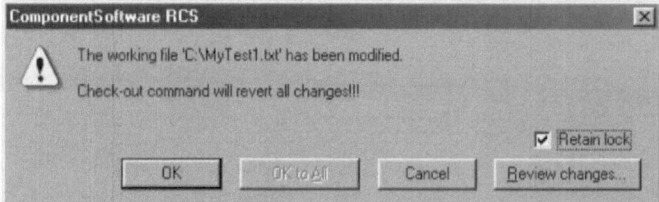

Figure 7.6

Ensure Retain lock is selected then select Review changes. The document is now back in its original state.

Step 6 *Checking in a new revision*
Modify the document MyText1.txt by adding and deleting a line. Right click the mouse button over the file in Windows Explorer and select Check-in. The following screen appears:

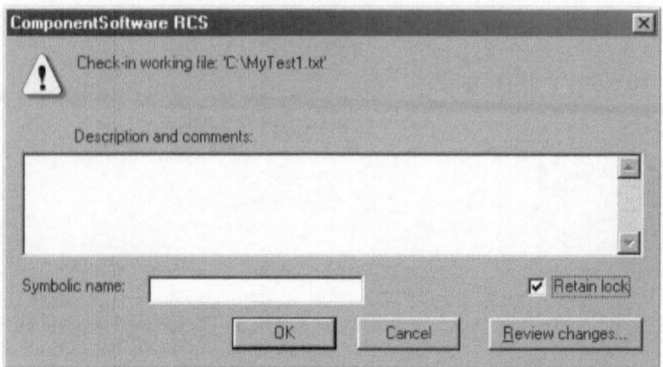

Figure 7.7

You can now enter a description about the changes and an appropriate change title. Then click OK.

Note: To allow other users to edit the file uncheck Retain lock.

Step 7 *Review the Revision History*
Right click on the file in MyTest1.txt in Windows Explorer. Select Revision History from the menu. The following page is displayed:

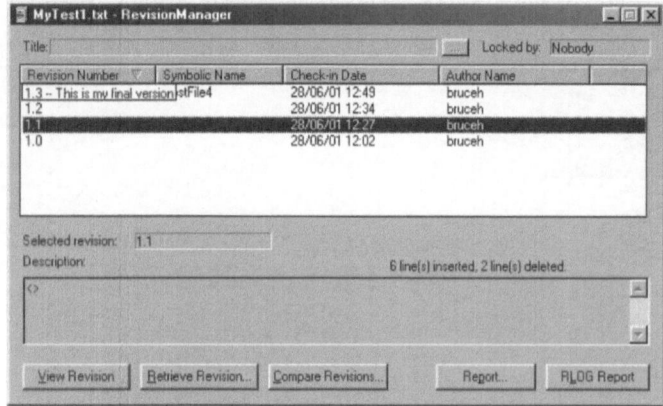

Figure 7.8

Several options are available for checking and comparing revisions.

Note: You can see the revision by double left clicking on the required version.

Step 8 *The Document Explorer*
The Document Explorer allows you to view and organize all documents that are checked into the Repository, as well as view general information about them. From the Document Explorer's Project menu, you can create projects, project milestones, branching, merging, and generate reports.

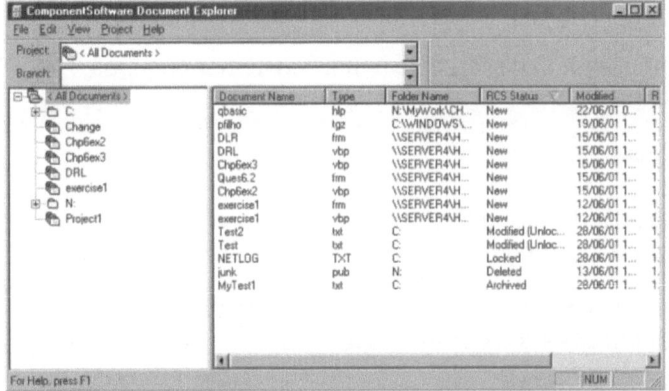

Figure 7.9

Software quality assurance – validation requirements

The *quality plan* would specify how validation requirements for a system are to be identified, expressed and accepted by the customer or end user. The fundamental concept of providing effective quality assurance testing techniques is to ask the question:

How can I develop a test that ensures that this particular requirement is met by the system that we are developing?

The validation requirements are formally extracted at the end of the requirements analysis phase of the development process, but they may be expressed informally at the start of the project.

Some basic points for validation requirements:

- Validation requirements are to be extracted from the system specification by the quality assurance staff assigned to the project
- The validation requirements suggested by the quality assurance staff must be discussed at the system specification review and between the development staff and the quality assurance staff
- Where possible, validation requirements and any test subsequently derived from them must be expressed in a decision table format (see below). Each row must contain the command being tested, the conditions under which it is being tested, for example the value of any relevant parameters, and the output expected from the test
- If the table format is not possible for any reason, then the validation requirements must be expressed in clear, concise and unambiguous English.

Below is an extract from a negotiated statement of requirements and a decision table:

The following command is provided for the signal operators
The CHECKSTATUS command. When the signal operator invokes this command the system checks that the line is clear and no trains are within a five mile radius of the junction. The system will respond with either a clear message (no trains in the area) or no clear warning signal that trains are in the vicinity of the crossing. The response must be immediate in case a train is approaching on the wrong track.

Command	Conditions			Output
	Command	Input signal	Response time	
CHECKSTATUS	correct	no trains	correct	correct – no trains message
CHECKSTATUS	correct	trains in area	incorrect	warning – response time to slow

A test plan in action

Generation of a test plan begins during the initial phases of a software project. The project manager needs to see the requirements of the testing process and estimate the resources required. When completing the requirements for a test plan the document needs to be detailed at a high level with the areas to be tested laid out in an abstract way. This will normally be sufficient for the management team to formulate a proposal and interact with a potential customer.

Below is an extract of a statement of requirements:

The Bernese Coach Company requires a computer-based system to handle customer enquires. The customer enquiries that need to be answered are:

- Give the travel price between any two destinations
- For a given route display the coach number
- Display the time each coach leaves for a specified destination
- Print the routes which are offering special party bookings.

A booking clerk is to operate the system by typing in one of the four commands, the command needs to be validated to ensure it is correct. The system uses a database called 'Coach Details' that contains all the information specified above.

Question
Carry out the following requirements:

- Specify an outline set of tests for the extract. The tests need to be structured so that the project manager can estimate the amount of testing resource required
- What type of resources would the project manager be able to predict from the information contained within a test plan?

Suggested solution
Function 1   Giving the travel price for any two destinations:

Test 1   Check the command input for the travel price to ensure that any incorrect syntax will be detected and a suitable error message displayed

Test 2   Check that incorrect destination details are handled with a suitable warning message. For example, if the wrong destination is entered, i.e. one that is on another route or one that the coach is not due to stop at

Test 3   Ensure that the travel price for the selected destinations is correct by manually checking against the latest printed fares

Test 4   Check to see that if one destination is not available (due to road works, diversions etc.) a suitable message is displayed along with the nearest alternative one.

Function 2   For a given route display the coach number:

Test 1   Check the command input for the coach number to ensure that any incorrect syntax will be detected and a suitable error message displayed

Test 2   Ensure that the selected route is available, if not a suitable message should be displayed

Test 3   Check that the coach number is correct for the required route against a printed chart

Test 4   Check that a suitable message is displayed if the coach number is not available for the selected route and suggest alternatives that might be available.

Function 3   Display the time each coach leaves for a specified destination:

Test 1   Check the command input for the timetable to ensure that any incorrect syntax will be detected and a suitable error message displayed

Test 2   Ensure that a suitable message is displayed if the destination is not correct or unavailable. If not available then an alternative destination is displayed (i.e. one close by)

Test 3   Check that the correct times are correct for selected destinations against the latest paper copy.

Function 4   Print the routes which are offering special party bookings:

Test 1   Check the command input for the special party bookings to ensure that any incorrect syntax will be detected and a suitable error message displayed

Test 2   Check the routes are correct against the latest printed information for the special party bookings

Test 3   Ensure that a suitable message is displayed if no special party bookings are available.

Note: These tests could be updated if the tender is accepted by the customer and the negotiated statement of requirements completed by the development team.

7.3 Project management

Introduction
Further information – 'Why projects go wrong'
The example below outlines some of the main causes of why projects do not meet their required stated objects:

> The National Health Service (NHS) decided to set up Information Systems across hospitals in the UK. It went ahead with one project despite high costs and risks. Even the suppliers did not think it was feasible. It chose three pilot projects within just two months. The hospitals selected were not prepared and no business mechanisms were established.

This example along with many other such cases are outlined in the Committee of Public Accounts Report titled 'Improving the Delivery of Government Projects' (ISBN 0 10 204700 6). The document includes information on 25 projects, the problems, the impact of the problems and the lessons learned.

On the issue of project management the report states that managers need more than a thorough grounding in the principles and practices, plus relevant experience. To be successful they need imagination and well-conceived risk management.

The 'lessons learned' outlined in the report are:

- Analyse and understand the full implications of proposed systems
- The specification must take into account the business needs and user requirements
- Look at the scale and complexity to assess if the project can be achieved
- Make sure senior management are involved
- High quality project management is essential
- Risk management and contingency plans are vital
- Develop close relationships with suppliers, but avoid undue reliance on them; retain ownership of progress
- Pay detailed attention to drawing up and managing contracts
- Plan in resources for staff training, do not underestimate this task
- Review projects so that lesson learned can be applied next time.

The report states that projects can be overtaken by technological changes because of delays. Senior management should be aware of the importance of halting a project that has been overtaken by events. An example is outlined below:

> A defence intelligence system was written off at a cost of £41m after it fell more than two years behind schedule and was overtaken by technology. The team did not recognize the complexity at the start, relied too much on contractors, without enough involvement, and should have introduced the system in phases.

In addition, project plans should be flexible enough to allow new technologies to be included where relevant. Organizations need to develop close relationships with suppliers in which they are able to share the same business vision.

Management structure
Figure 7.10 outlines a typical management structure.

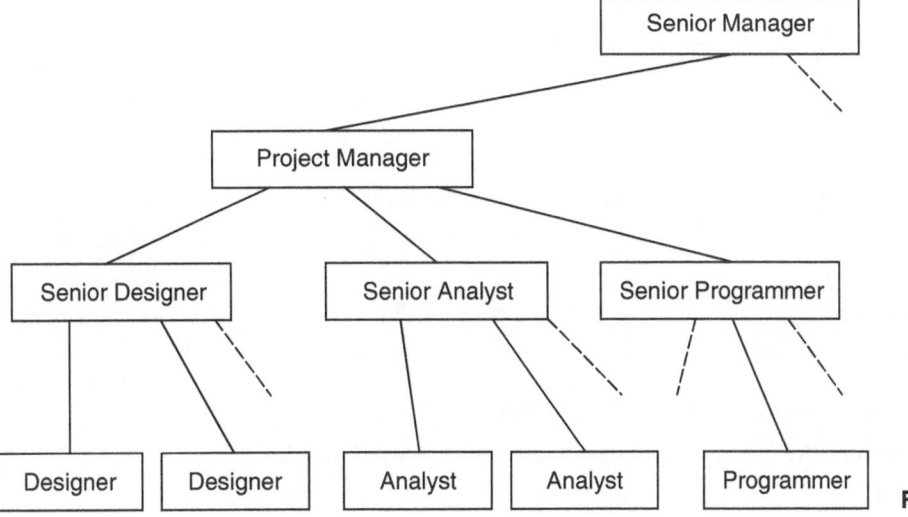

Figure 7.10

A quote about management:

Management must always, in every decision and action, put economic performance first. It can only justify its existence and its authority by the economic results it produces. There must be great non-economic results: the happiness of the members of the enterprise, the contribution to the welfare of culture of the community, etc. Yet management has failed if it fails to produce economic results. It has failed if it does not improve or at least maintain the wealth producing capacity of the economic resources entrusted to it. The first definition of management is therefore that it is an economic organ of an industrial society. Every act, every decision, every deliberation of management has as its first dimension and economic dimension.

(*Drucker*)

Question
Is the above quote appropriate when a manager is handling an organization like a charity?

Answer
Yes: charities have to make money for their respective organizations, i.e. OXFAM.

Solution to Exercise 7.3.1
(i)

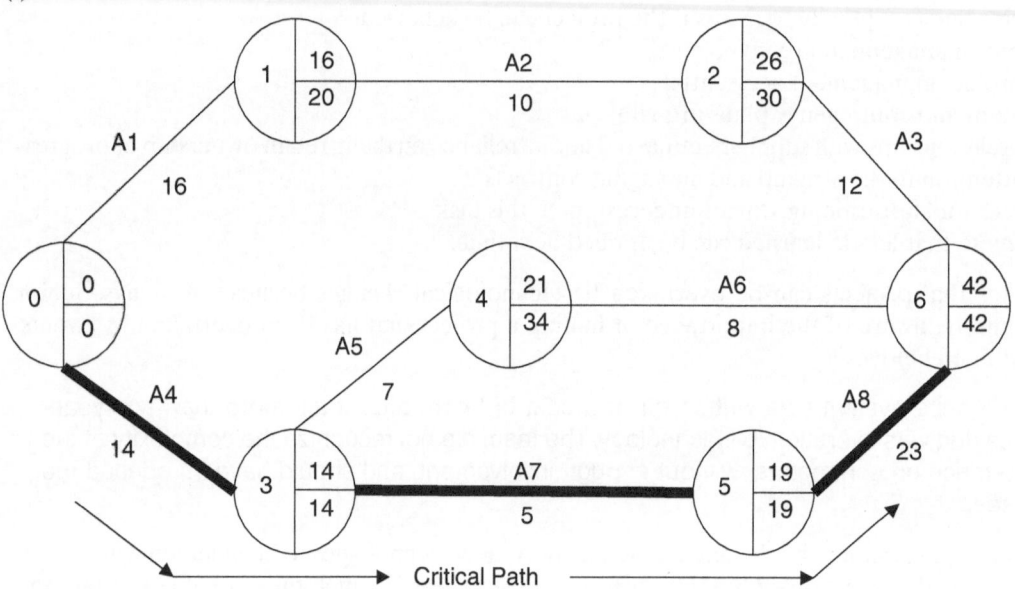

Figure 7.11 *Outline network diagram*

(ii)

Activity	Estimated duration	Maximum span	Total float
A1	16	20	4
A2	10	14	4
A3	12	16	4
A4	14	14	0
A5	7	20	13
A6	8	21	13
A7	5	5	0
A8	23	23	0

(iii)
The critical path lies along the line where the total float is zero.

Therefore the critical path lies along the line A4, A7 and A8.

This is outlined in Figure 7.11.

Critical path analysis – additional exercise

Figure 7.12 represents a series of software development activities that are all carried out by a single software engineer. The duration in days is included on the diagram.

(i)   Complete the missing Earliest Estimated Times (EET) and Latest Allowable Times (LAT) by filling in the missing spaces on the diagram
(ii) Create a table that specifies the maximum span and total floats and from this specify the critical path
(iii) The activities in the diagram are now to be rescheduled to a team of software engineers. Activities A1, A2 and A3 are allocated to software engineer 1, activities A4, A8 and A10 are allocated to software engineer 2, activities A6, A7 and A5 are allocated to software engineer 3 (all these activities start work on day 0). A fourth software engineer is to be brought in at the earliest possible date to undertake activity A9. Prepare a Gantt chart to represent this schedule. Can all ten activities be completed by fewer than four software engineers?

Problem

(i)

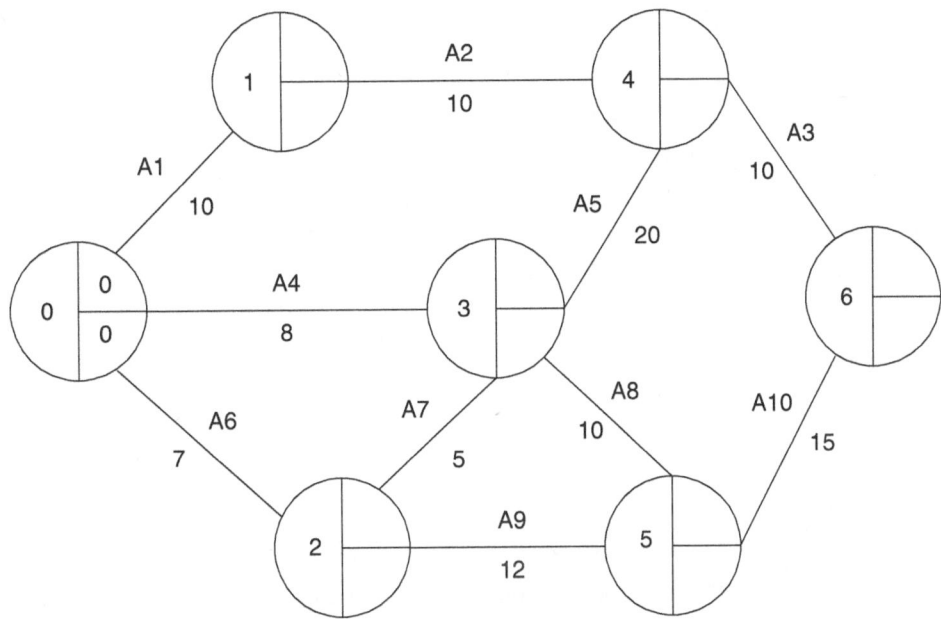

Figure 7.12

Solution

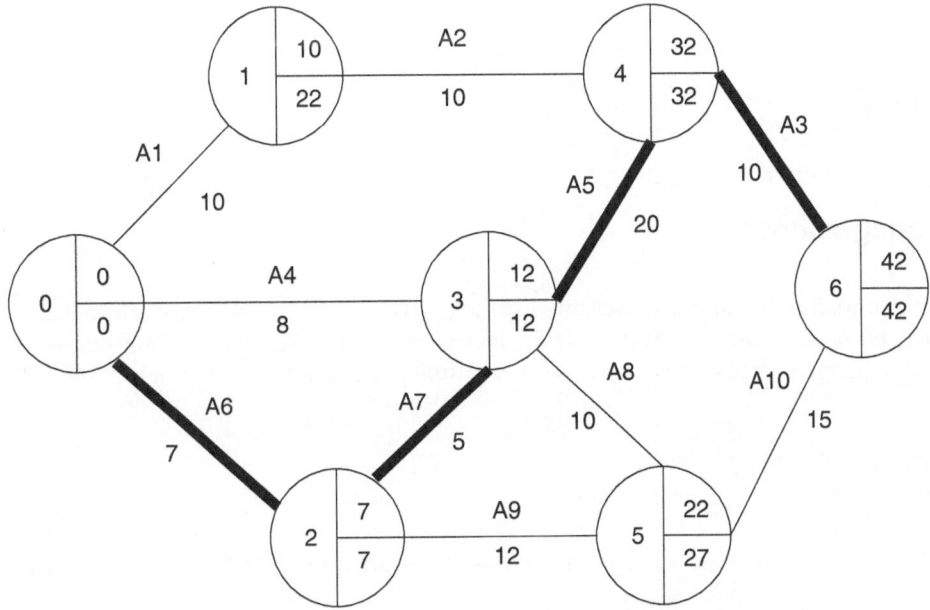

Figure 7.13

Additional exercise (further solutions)

(ii)

Activity	Estimated duration	Maximum span	Total float
A1	10	22	12
A2	10	22	12
A3	10	10	0
A4	8	12	4
A5	20	20	0
A6	7	7	0
A7	5	5	0
A8	10	15	5
A9	12	20	8
A10	15	20	5

Activities with a total float of 0 identify the critical path, in this case A6, A7, A5 and A3.

(iii)

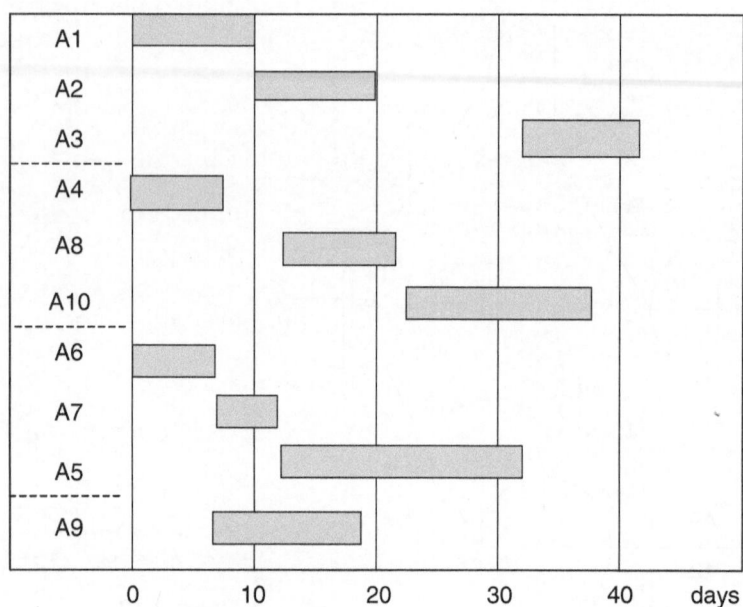

Figure 7.14

The critical path has 42 days. The total programming time is 107 days so three software engineers could complete the task with 19 slack days. See the table below:

Staff	Activities	Free days
Software engineer 1	A1, A2, A8	with 12 days free
Software engineer 2	A4, A9, A10	with 7 days free
Software engineer 3	A6, A7, A5, A3	no free days

Tools for project management

Microsoft Project

Microsoft Project 2000 is a powerful, flexible tool designed to help you manage a full range of projects. Schedule and closely track all tasks – and use Microsoft Project Central, the Web-based companion to Microsoft Project 2000, to exchange project information with your team and senior management.

Some main features:

IMPROVE TEAM PRODUCTIVITY
1. Collaborative planning with microsoft project central
   Personal Gantt chart
   Render Gantt views like those in Microsoft Project to outline each team member's own tasks across multiple projects.

New task

Team members can create tasks, which a project manager can then approve before adding them to the project plan.

Task delegation

Once assigned by the project manager, tasks may be delegated from team leaders to team members or from peer to peer. The Delegation feature can also be disabled if desired.

Show Microsoft Outlook tasks

Team members can display entries from their task list in the Microsoft Outlook® messaging and collaboration client so they can see all their project and non-project tasks in a single location.

View non-working time

Team members can report non-working time to the project manager, such as vacation or sick leave, and also report work time that cannot be devoted to the project.

Workgroup

Project managers can assign task responsibilities and track project status across workgroups to keep the project on track.

2. Collaborative tracking with Microsoft Project Central

    Auto-accept rules

    Managers can reduce time spent in administrative activities by establishing rules to automatically accept actual hours, percentage completed, or any information in a custom field.

    Status reports

    Create custom report formats and request or receive team member status updates, which Microsoft Project Central automatically rolls into a group report.

    Administration module

    Project administrators can gain more control over definitions of non-project time, views, formatting styles, and security to ensure consistency in management approaches and organization structure.

    Timesheet

    Team members can see their assignments across projects, enter updates, and easily send them to the project manager.

    Actual hours and per cent complete tracking

    Team members can track and report actual hours spent on each assigned task, or estimate what percentage of the task is complete when tracking is difficult or deemed unnecessary.

3. Easy access to project information

    Views

    Senior executives, managers, and team members can access different views of projects, such as View Your Portfolio, View Your Project, and View Assignments.

    Offline capabilities

    Team members can take their timesheets and status reports offline and continue working on them from any location.

INCREASE DATA USEFULNESS

1. Flexible viewing

    Grouping

    Quickly categorize and view task and resource information in any grouping that is most useful to you.

    Outline codes

    Define outline codes instead of having to tie them to the outline structure of a project.

    Graphical indicators

    Associate graphical indicators with the data in a custom field, so a particular image can be displayed in place of the actual data to easily spot potential problems.

    Fiscal year in timescale

    Independently set the use of the fiscal year for both major and minor timescales to display data in a specific timescale combination.

    Network diagram

    Customize network diagrams with new filtering and layout options, increased formatting features, and enhanced box styles (formerly the PERT chart).

    Rollup Gantt

    Display Gantt bars for all subtasks on a single task summary line.

2. Flexible analysis

Task calendars

Create schedules with task-specific calendars that affect only selected tasks.

Materials resources

Specify consumable resources such as lumber or concrete and assign them to tasks.

Deadline dates

Remind your team of deadlines and alert them visually if deadlines cannot be met.

Cross-project critical path

Calculate the critical path within individual or across all inserted projects to see a single critical path for the overall master project.

Custom fields: value lists

Define pick-lists to restrict the values that can be entered into a custom field and simplify data entry with just a mouse click.

Custom fields: formulas

Add custom formulas for arithmetic calculations, conditional testing, and functions to be applied to custom field data.

OLE DB

Share Microsoft Project 2000 data with other applications and integrate data across the enterprise using OLE DB.

Estimated durations

Indicate a tentative duration for a given task by simply entering a value for the duration followed by a question mark; then return to the task at a later time to enter a confirmed duration.

Month duration

Microsoft Project now supports months as a unit of duration.

Contoured resource availability

Create plans that incorporate time-phased resource availability information.

Clear baseline

Clear the baseline or interim plan data for selected tasks or an entire project.

3. Easier reporting

Scaling and printing

Print documents more efficiently and easily with new and improved printing and scaling options.

Copy picture

Create higher quality images with larger allowable sizes and better scaling.

4. Enhanced user confidence

HTML Help

Enables easier interaction between the procedures found in HTML Help, Microsoft's new standard help system, and your work in Microsoft Project 2000.

Adaptive menus

Just as in Office 2000, only the items that you use most often are prominently featured on the menu.

Templates

Create and access Microsoft Project templates easily.

Variable row height

Drag the row line between tasks to set individual rows to the height you want.

In-cell editing

View the context of a task while editing it.

Fill Handle

Select the cells you want to populate, and use the Fill Handle feature to easily copy information or repeat patterns in several locations.

AutoSave

Set Microsoft Project 2000 to automatically save work at chosen time intervals to avoid losing valuable data if your computer is inadvertently shut down.

Office server extensions support

Save to a Web server just as easily as you save to network locations.

Single Document Interface

Similar to Office 2000 applications, Microsoft Project 2000 supports the Single Document Interface.

Default save path and format

Specify a default save path and format to save project data where and how you want it.

Accessibility

Get support for third-party accessibility aids through Microsoft Active Accessibility programming interfaces.

Project Open and Save

The Open and Save dialog boxes now feature the same look and functionality as those in Office 2000.

Hyperlinks

Easily link to frequently browsed Web pages or commonly used files with the new Office 2000 Hyperlink dialog box.

EXTEND PROJECT MANAGEMENT

5. Improve capacity and performance

    Database performance

    Get improved performance and access to data with changes to the Microsoft Project database.

    Resource pooling

    Improve network performance by using a resource pool.

    Inserted projects

    If you ever need to move a project, all the links from master projects to subprojects will automatically be maintained.

6. Keep users up and running

    Install on demand

    Install only the components you need – when you need them.

    Roaming user support

    Enhanced portability lets you log on to any computer in a networked environment and maintain personal settings and preferences.

    Windows terminal server support

    Microsoft Project 2000 can now run on Microsoft Windows® Object Model.

    The outline shows a typical Gantt chart display with a list of associated tasks.

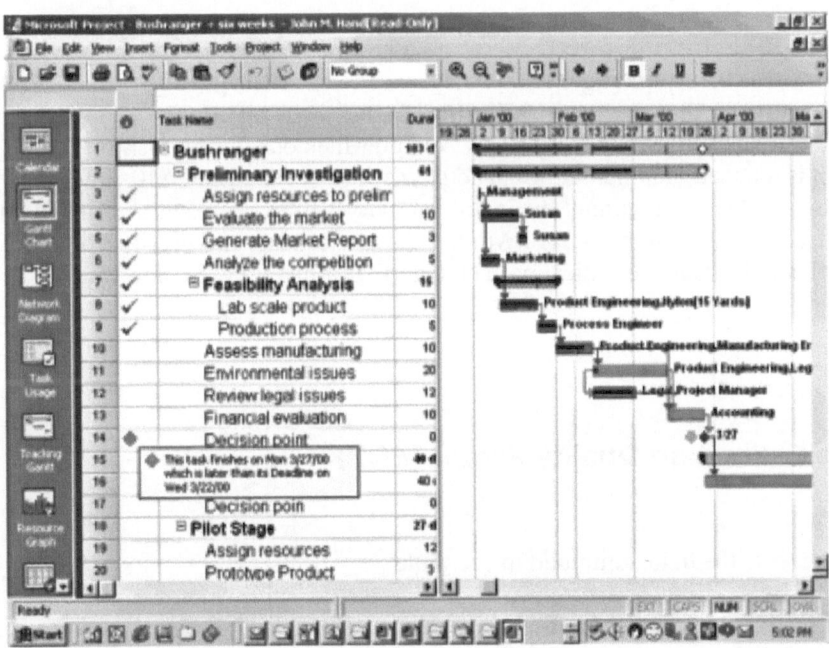

Figure 7.15 *Microsoft Project 2000 pictorial outline*

Microsoft Project books

Microsoft Project 2000 Step by Step with CD-ROM by Carl S. Chatfield and Timothy D. Johnson. Published by Microsoft Press

Using Microsoft Project 2000 by Tim Pyron. Published by Que

Microsoft Project 2000 for Dummies by Martin Doucette. Published by International Data Group, Incorporated

Effective Executive's Guide to Project 2000: The Eight Steps for Using Microsoft Project 2000 to Organize, Manage and Finish Critically Important Projects by Stephen L. Nelson, Pat Coleman and Kaarin Dolliver. Product code (ISBN): 0967298113

Further information

A lot of additional information, including tutorials, downloads, trial software, case studies, training and certification, support etc. can be found at:

http://www.microsoft.com/office/project/default.htm

Additional example using Microsoft Project

Below is a list of activities, their durations and dependencies:

Activity (task)	Estimated duration of the activity (in days)	Dependent
A1	10	–
A2	10	A1
A3	10	A2, A5
A4	5	–
A5	20	A4, A6
A6	8	–
A7	12	A6
A8	10	A4, A6
A9	15	A7, A8

This table is to be documented within a project management software tool.

(i)   Open up the software tool (i.e. Microsoft Project) and create a new project
(ii) Enter the tasks above to form a corresponding Gantt chart
(iii) Display the corresponding network diagram, state the critical path
(iv) Add the following two milestones (Milestone 1 after task A4 and Milestone 2 after task A7)
(v)   Before the project starts it is realized that an additional activity, which we shall call A10, is going to be needed. A10 has a duration of two days. It cannot begin until activity A1 has been completed and must itself be completed before activity A5 can be started. Modify the Gantt chart and network diagram to incorporate the additional activity. Has the critical path changed? If it has state the new one.

Suggested solution

See Figures 7.16–7.20.

Further example on the Estimation Quality Factor (EQF)

From the table below:

(a) Complete a graph that represents the data contained in the table
(b) From the graph calculate the EQF
(c) Comment on the results (i.e. are they acceptable or not?).

Estimate number	Date	Cost of estimate (£m)
1	1-4-2000	2.25
2	1-7-2000	2.50
3	15-10-2000	4.25
4	1-2-2001	4.00
Actual	1-4-2001	3.75

Gantt chart

(ii)

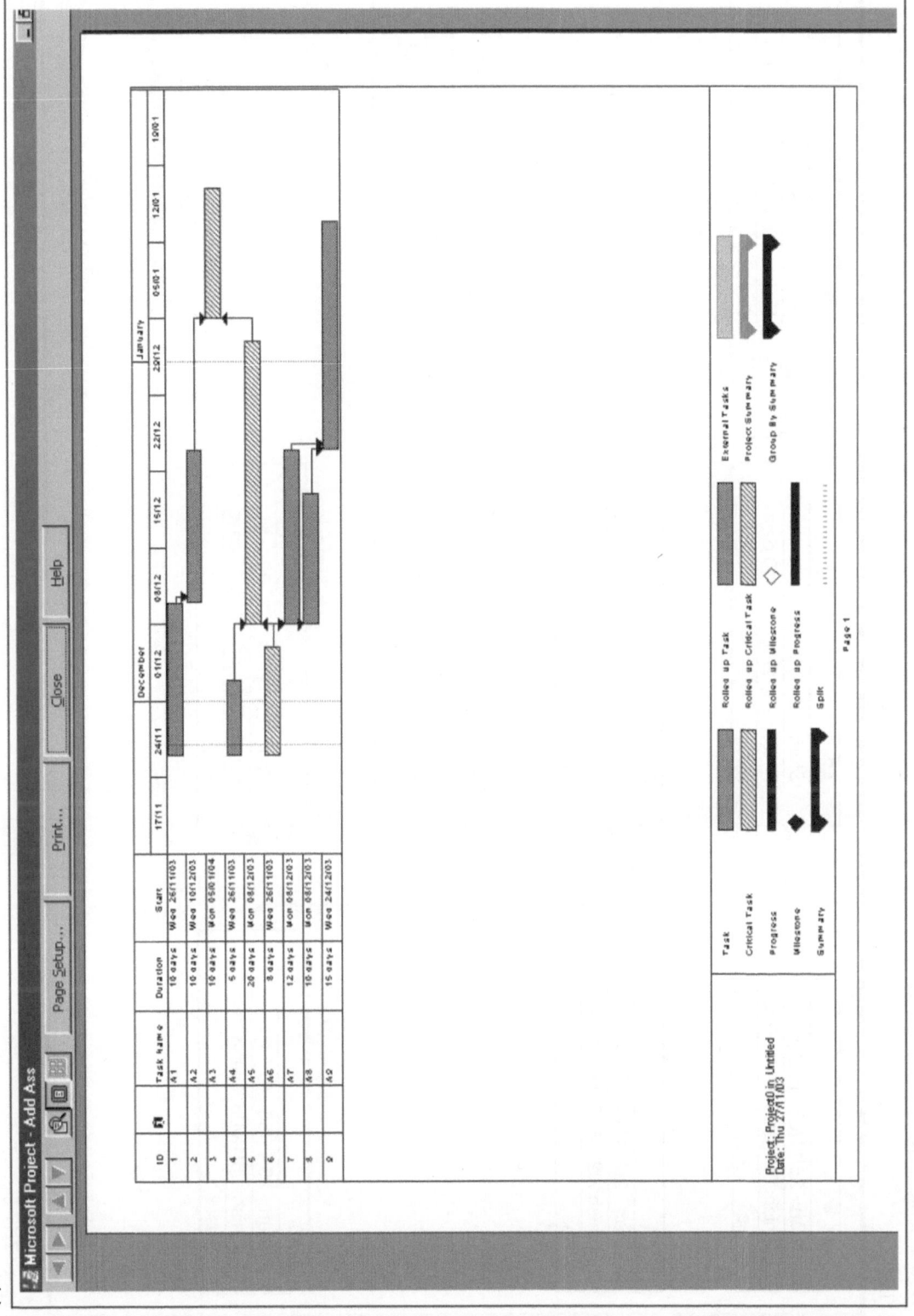

Figure 7.16

Network diagram

(iii)

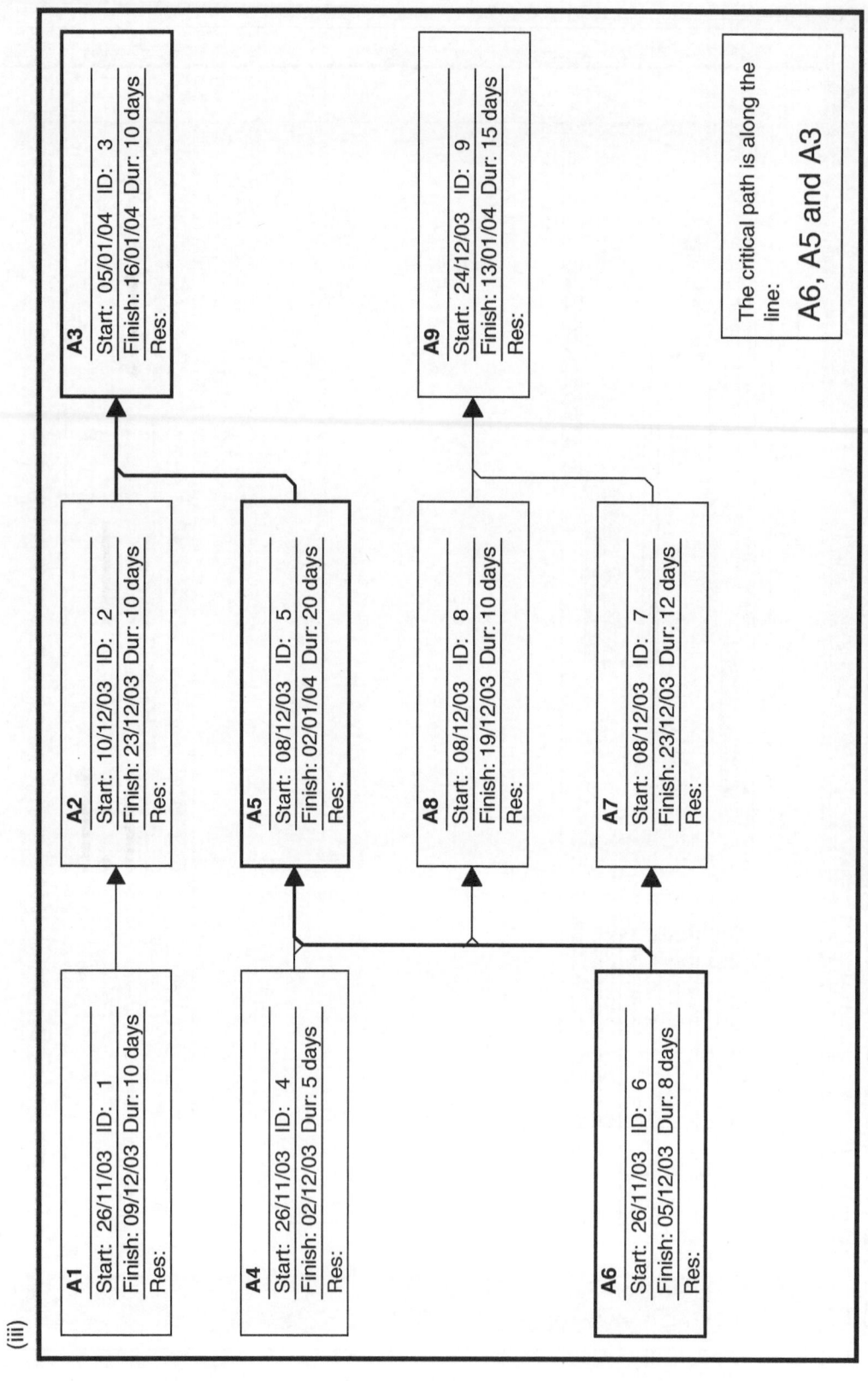

A1		A2		A3
Start: 26/11/03 ID: 1		Start: 10/12/03 ID: 2		Start: 05/01/04 ID: 3
Finish: 09/12/03 Dur: 10 days		Finish: 23/12/03 Dur: 10 days		Finish: 16/01/04 Dur: 10 days
Res:		Res:		Res:

A4		A5		A9
Start: 26/11/03 ID: 4		Start: 08/12/03 ID: 5		Start: 24/12/03 ID: 9
Finish: 02/12/03 Dur: 5 days		Finish: 02/01/04 Dur: 20 days		Finish: 13/01/04 Dur: 15 days
Res:		Res:		Res:

	A8
	Start: 08/12/03 ID: 8
	Finish: 19/12/03 Dur: 10 days
	Res:

A6		A7
Start: 26/11/03 ID: 6		Start: 08/12/03 ID: 7
Finish: 05/12/03 Dur: 8 days		Finish: 23/12/03 Dur: 12 days
Res:		Res:

The critical path is along the line:
A6, A5 and A3

Figure 7.17

Milestone additions

(iv)

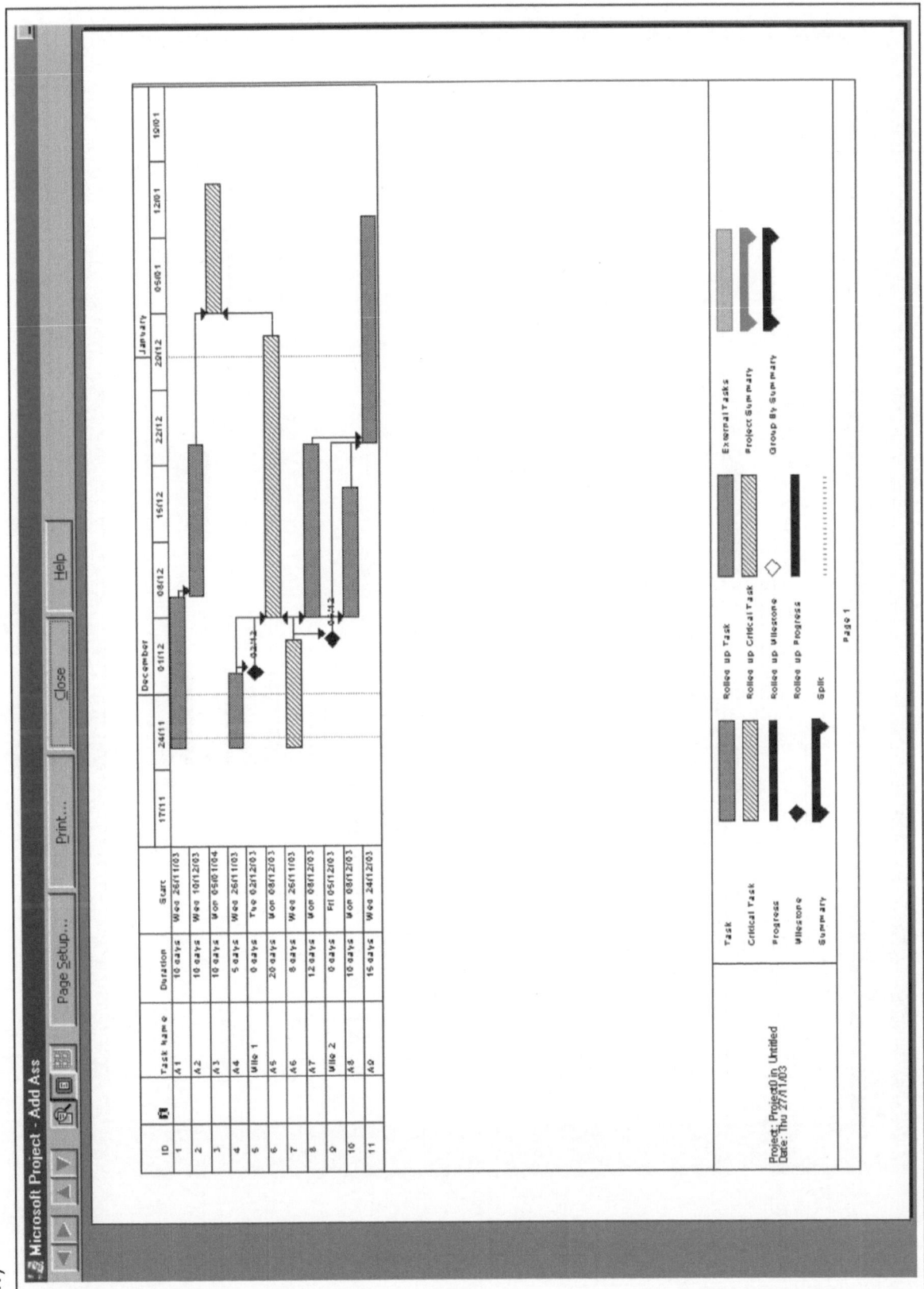

Figure 7.18

Modifications

(v)

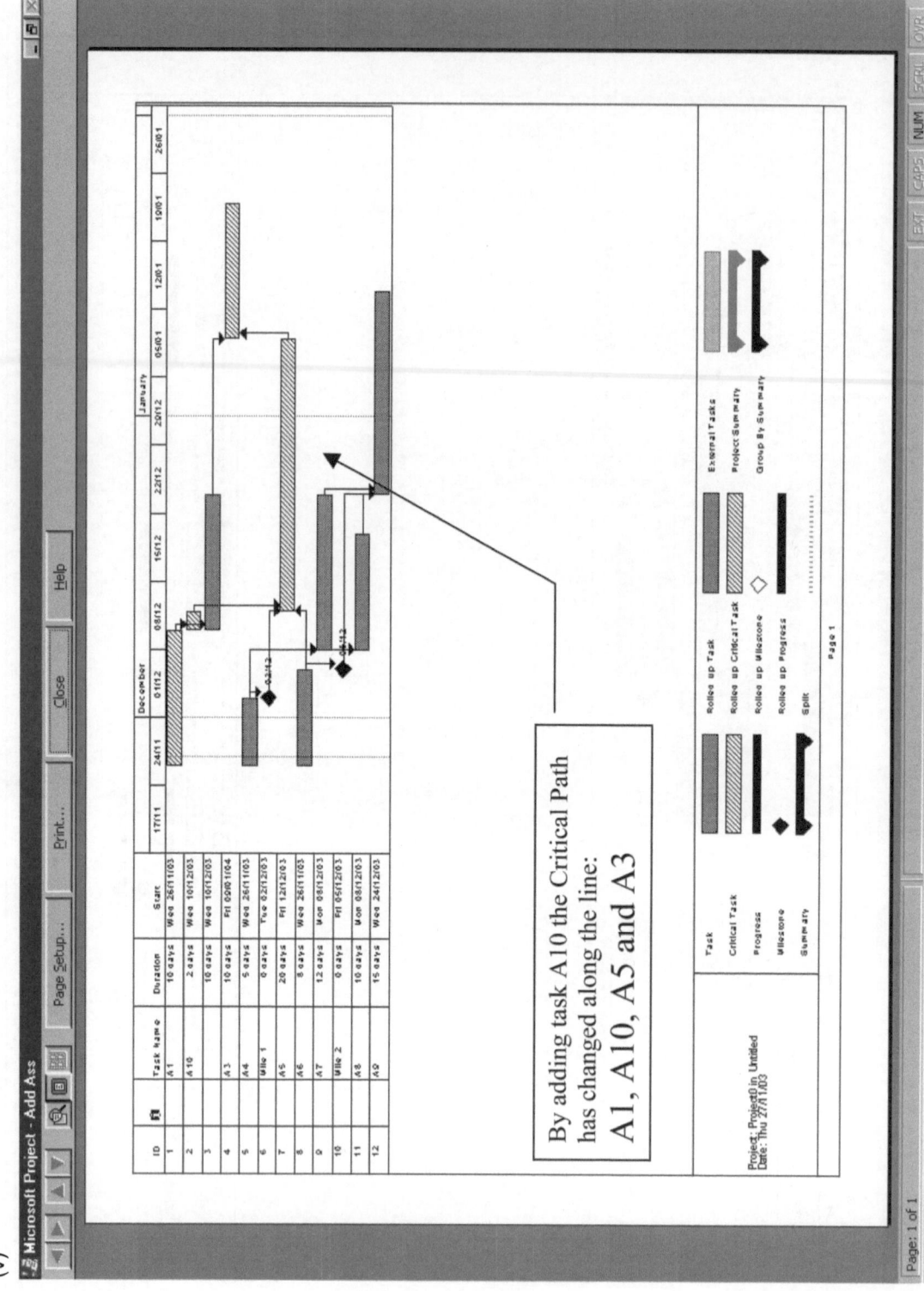

By adding task A10 the Critical Path
has changed along the line:
A1, A10, A5 and A3

Figure 7.19

Network diagram

(vi)

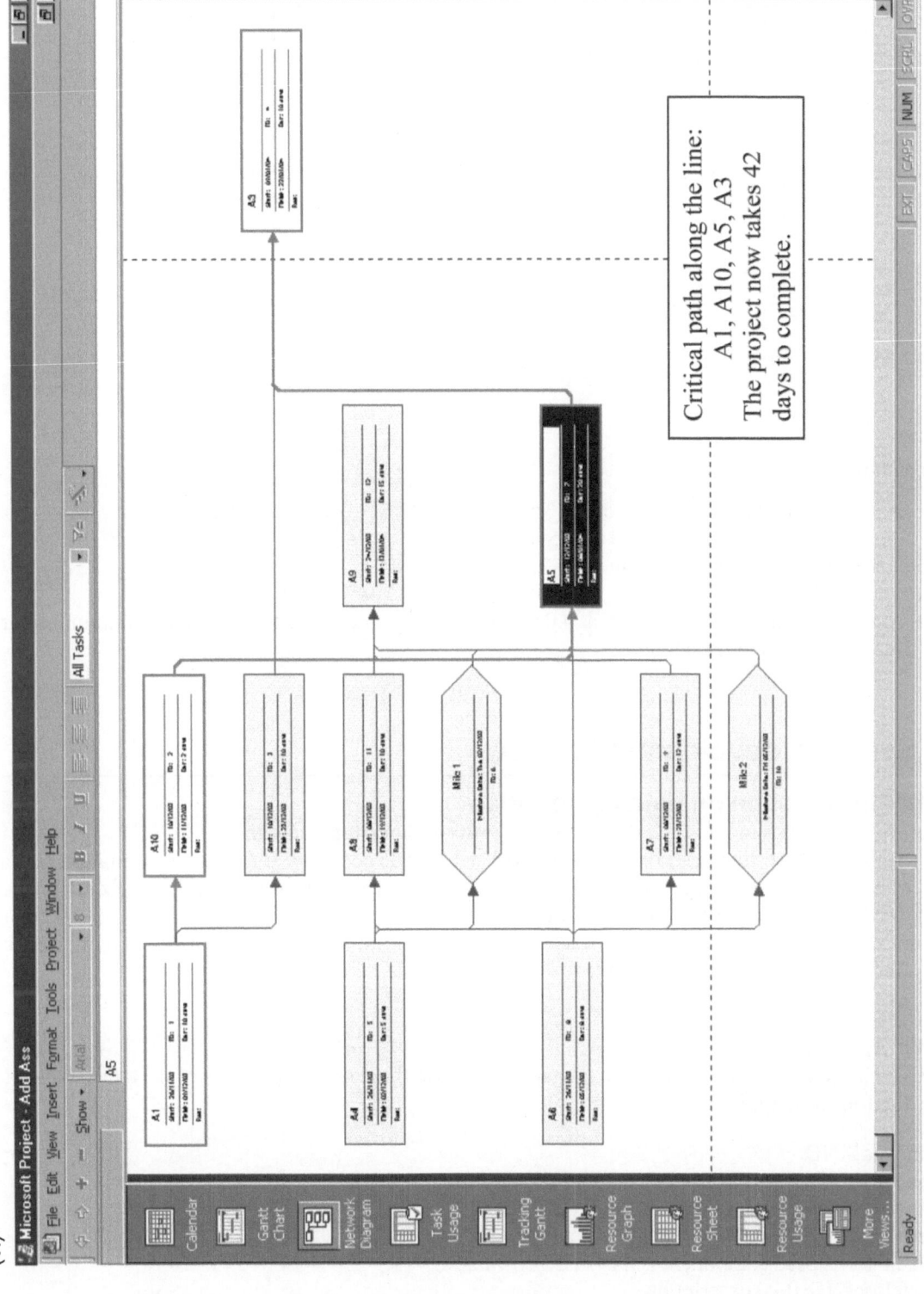

Figure 7.20

Suggested solution
(a) The graph

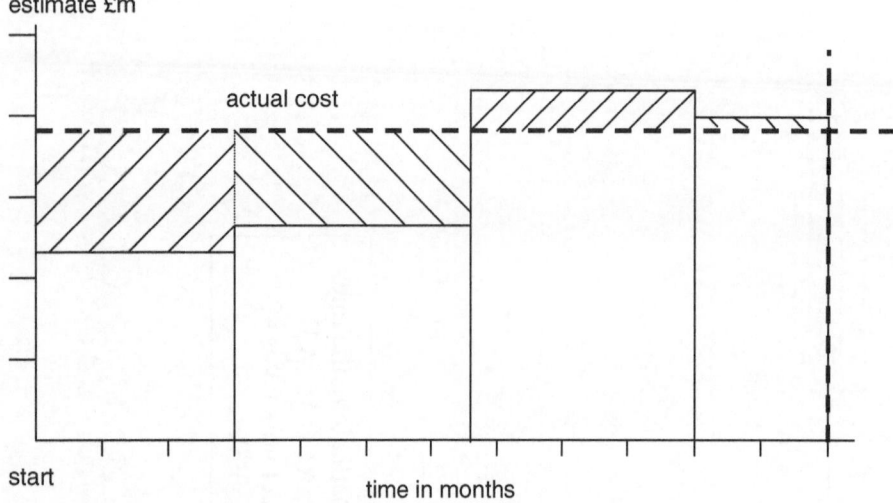

estimate £m

start

time in months

Figure 7.21

(b) Calculation:

Area of the shaded area: $(3 \times 1.5) + (3.5 \times 1.25) + (3.5 \times 0.5) + (2 \times 0.25)$
$= 4.5 + 4.375 + 1.75 + 0.5 \qquad = 11.125$

Total area under the actual cost line $= 12 \times 3.75 \quad = 45$

Therefore EQF $= 45/11.125 \quad = 4.045$

(c) This is only just an acceptable EQF as 4.045 is only just greater than 4 which is the minimum figure set by DeMarco as a quality standard.

7.4 Systems development review

Further information on data dictionaries
Usage: Within analysis models like data flow and entity-relationship diagrams.

Generally these are computer based and provide a central repository for all the information contained within a modelling process. They are used to provide automatic analysis and checking of entries that are applied in the development of a proposed model.

From the System Analysis Unit we can see that a special notation is used within a data dictionary based on the BNF (Backus-Naur Format) syntax:

@BNF syntax
The BNF clause is used in the data dictionary to describe the data components of data flows and stores. It uses the following syntax:

+   means AND
[|]   means EITHER-OR, i.e. select one of the options
( )   means enclosed component is OPTIONAL
{ }   means ITERATIONS OF enclosed component
" "   means enclosed component is a LITERAL value
\\   encloses a textual description

The iteration braces { } are often annotated with lower/upper limits. For example:

3{dog}7   between 3 and 7 iterations of dog
1{cat}   one or more iterations of cat
{horse}3   up to 3 iterations of horse

The BNF expression can contain a mix of these operators:

@BNF = [lion | cat + 2{dog}2 + (horse) | danger]

Further examples:

@NAME = MESSAGE
@TYPE = Discrete Flow
@BNF = \ single character A-Z in either upper or lowercase \
@NAME = OCTAL DIGIT
@TYPE = Discrete Flow
@BNF = ["0"|"1"|"2"|"3"|"4"|"5"|"6"|"7"]

Note that the '+' sign should not be confused with the arithmetic use of the symbol. It is used to represent a combination of data components into a composite structure. It can be considered as an 'and' operation with the '[|]' taking the 'or' equivalent.

Be a little careful with the 'or' statement as the whole expression that is to be included in the comparison must be included inside the square brackets, i.e. [aFlow | bFlow | cFlow] is an inclusive or statement based on the three flows.

Data flow diagram example

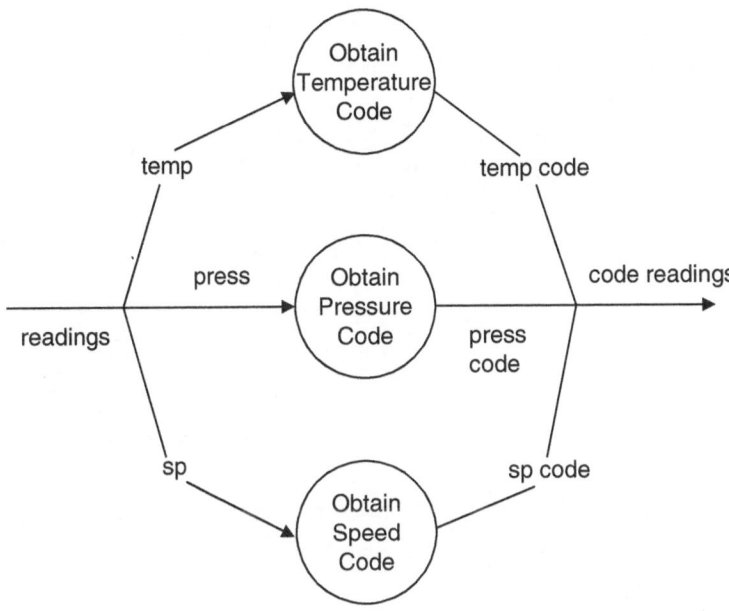

Figure 7.22

Figure 7.22 shows two group nodes that need to be specified using correct BNF notation. A data dictionary for the 'readings' data flow is outlined below.

Data Flow: readings
       readings = temperature + pressure + speed
Notes: all the readings need to be converted in order to produce the required code settings
Alias:

temperature =
       * temperature in Celsius expressed as a real number *
Notes:
Alias: temp

pressure =
       * pressure in Pascals expressed as a real number *
Notes:
Alias: press

speed =
       * speed in revs/min expressed as a real number *
Notes:
Alias: sp

Looking at the Yourdon Select CASE tool package (using the Bess and Bailey Kennel System example), the dictionary shows the components as shown in Figure 7.23.

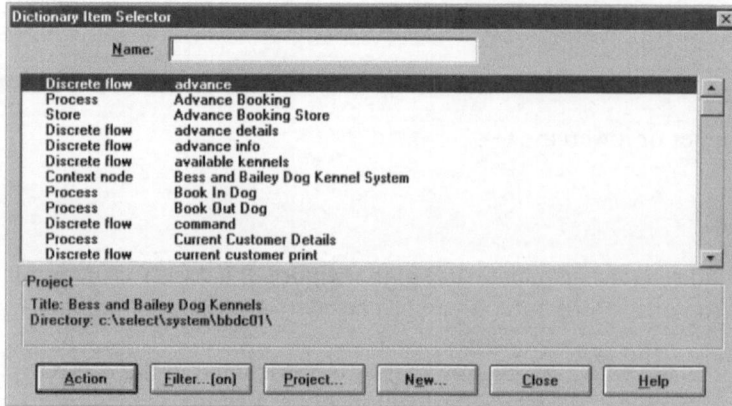

Figure 7.23

Looking at the valid command data flow (group node input) and selecting edit results in Figure 7.24.

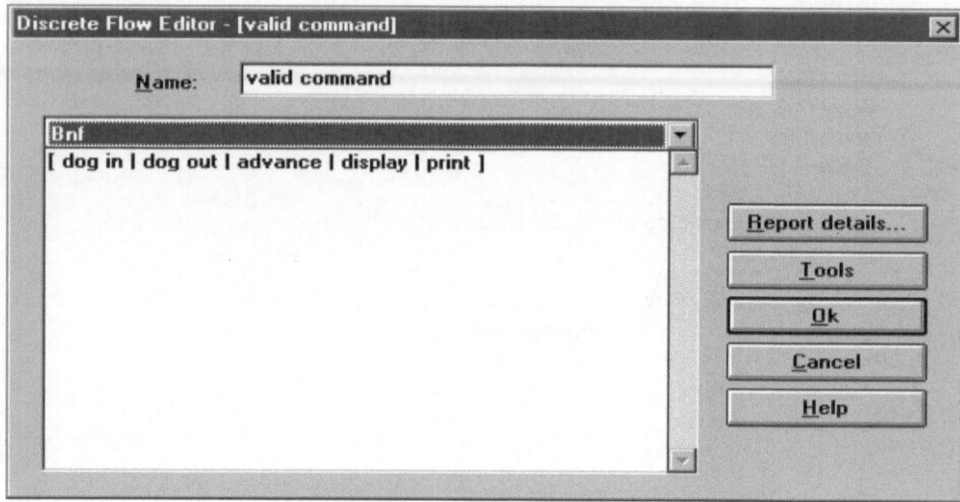

Figure 7.24

Which shows the appropriate BNF syntax (inclusive 'or' outputs) and the report contains:

Project: C:\SELECT\SYSTEM\BBDC01\
Title: Bess and Bailey Dog Kennels
Date: 10-Apr-2001 Time: 13:4

Name: **valid command**
Type: Discrete flow
Bnf: [dog in | dog out | advance | display | print]
This item is used on the following diagrams:
    DOG2.DAT   Bess and Bailey Dog Kennel System
    DOG3.DAT   Process Receptionist Command
Last changed: DEFAULT   19-Dec-2000 19:06:01

---- End of report ----

Note: To add comments and other information like appropriate units select the options arrows to the right of the BNF term (see Figure 7.25).

The data dictionary can be used to express more precise information, for example the field structures within the stores and the maximum number of records to be held.

Data dictionaries – additional student exercise

Produce data dictionary entries for the data described in the statement of requirements extract outlined below:

The system needs to display personal information about current students. This should include full name, address, date of birth and telephone number. Students can be classified as under-graduates

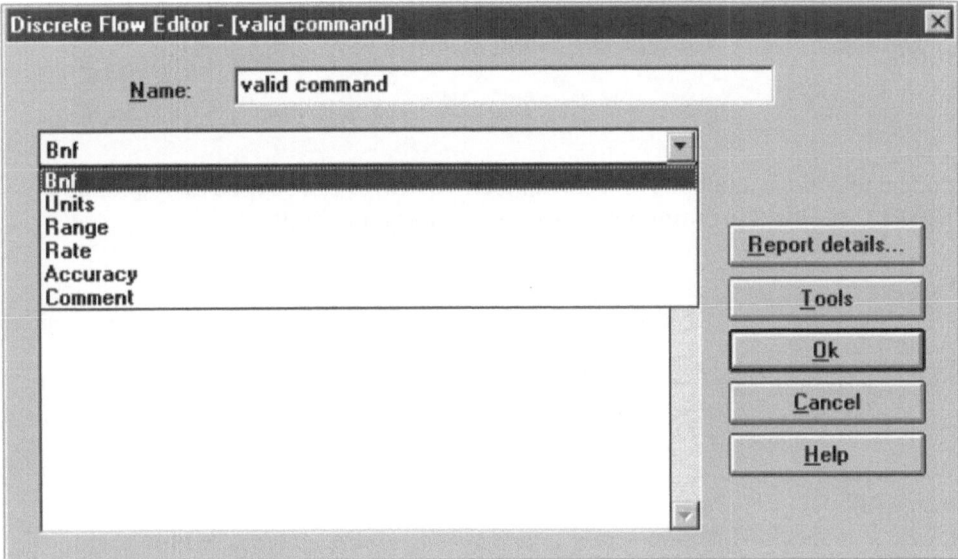

Figure 7.25

or post-graduates. The system also needs to determine the next of kin to include their names, addresses and telephone numbers.

The following definitions are to apply:

A name is composed of 1..20 characters, i.e. 1{ Character } 20.
An address is composed of a house number or name, street, town, county and a postcode.
The date-of-birth is made up of day (2 digit number), month (2 digit number) and year (4 digit number), i.e. for somebody born on 13 March 1979 the actual date requirement is: 13 03 1979.
The telephone numbers are to be between 8 and 11 digits to cater for local and non-local students and next-of-kin, i.e. 8 { Digit } 11.
Student classification will be composed of 14..15 characters, where 14 characters are required for a post-graduate student and 15 characters are required for an under-graduate student.

Complete a data dictionary to model the above requirements.

Suggested solution

Data Flow: Student
Student =
         Student Name + Student Address + Date Of Birth + Student Telephone + [Under-Graduate | Post-Graduate] + Next Of Kin Name + Next Of Kin Address + Next Of Kin Telephone
Notes: Required student details to be displayed
Alias:

Student Name:
         1 { Character } 20
Notes: Current student name
Alias: SName

Student Address:
         [House Number | House Name] + Street + Town + County + Post Code
Notes: Current student address
Alias: SAddress

Date Of Birth:
         2 { Digit } 2 + 2 { Digit } 2 + 4 { Digit } 4

Notes: Student's date of birth expressed as a day, month and year, i.e. 12 05 1981
Alias: DateOfBirth

Student Telephone:
        8 { Digit } 11
Notes: Current student telephone number. For local numbers 8345 6789 can be used as 8 continuous
        digits or 020 8322 1111 as 11 continuous digits for non-local numbers
Alias: SPhone

Under-Graduate:
        15 { Character } 15
Notes: Under-Graduate student classification
Alias: UGrad

Post-Graduate:
        14 { Character } 14
Notes: Post-Graduate student classification
Alias: PGrad

Next Of Kin Name:
        1 { Character } 20
Notes: Emergency contact name for the next of kin to be informed in an emergency
Alias: NOKName

Next Of Kin Address:
        [House Number | House Name] + Street + Town + County + Post Code
Notes: Address of the next of kin
Alias: NOKAddress

Next Of Kin Telephone:
        8 { Digit } 11
Notes: Emergency telephone number for the next of kin
Alias: NOKPhone

Control aspects

Suggested solution to Exercise 7.4.1 – Bailey Comfortable Room Company

Context diagram

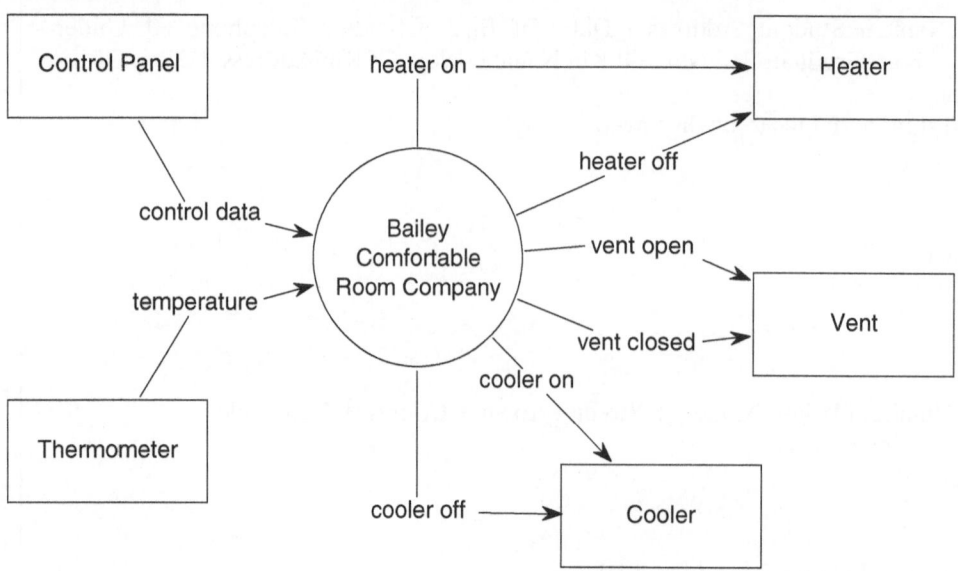

Figure 7.26 *Bailey Comfortable Room System: context diagram*

DFD refinement

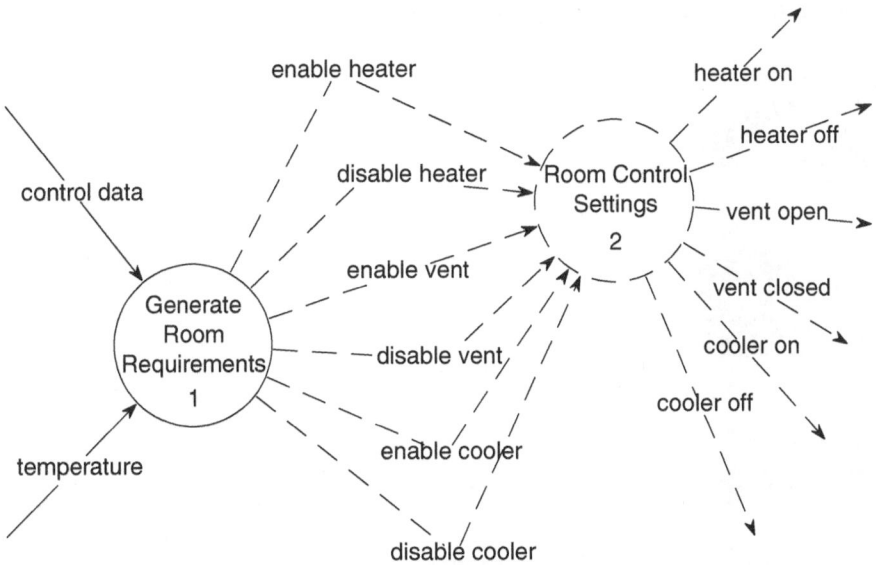

Figure 7.27 *Bailey Comfortable Room Company: DFD refinement*

Remember a state transition diagram is a child of a control process

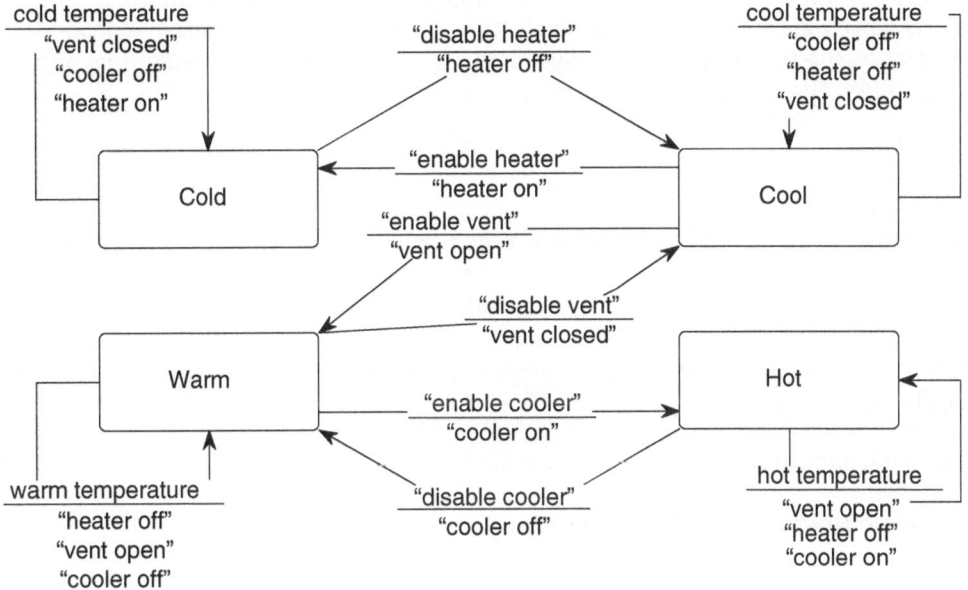

Figure 7.28 *Room control settings*

Syntax consistency check

Project: C:\SELECT\SYSTEM\ROOM1\
Title: Bailey Comfortable Room Company
Date: 17-Apr-2001 Time: 17:1

Report: Diagram Consistency checking

This report contains a consistency check of all the diagrams in the project.

Checking ROOM1.DAT
    No Errors detected, No Warnings given.

Checking ROOM2.DAT
    No Errors detected, No Warnings given.

Checking ROOM3.DAT
    No Errors detected, No Warnings given.

---- End of report ----

Additional example – control aspects
Susie South West Train Company train level crossing control system

Figure 7.29

Below is part of a specification for the Susie South West Train Company.

> The Susie South West Train Company requires a computerized control system for one of its single-track railway routes. The system is to control an unmanned level crossing point where the railway crosses a main road. A sensor will detect the approach of a train which first starts the warning lights flashing and then drops down the barriers. The crossing is now closed to all road vehicles. A second sensor will then detect when the train is leaving the level crossing which first opens the barriers and then stops the lights flashing. The crossing is now open to all road vehicles.

From the above specification:

1. Complete a data flow context diagram
2. Complete a refinement for the context diagram to include the required control process and associated control flows
3. Create a child of the control process to produce a state transition diagram showing the dynamic aspects of the level crossing system
4. Produce a consistency test to ensure that the correct parent–child balancing is maintained and the diagrams are syntactically correct.

Note: You are to complete the above requirements using a suitable CASE tool application.

Suggested solution – Susie South West Train Company

Context diagram

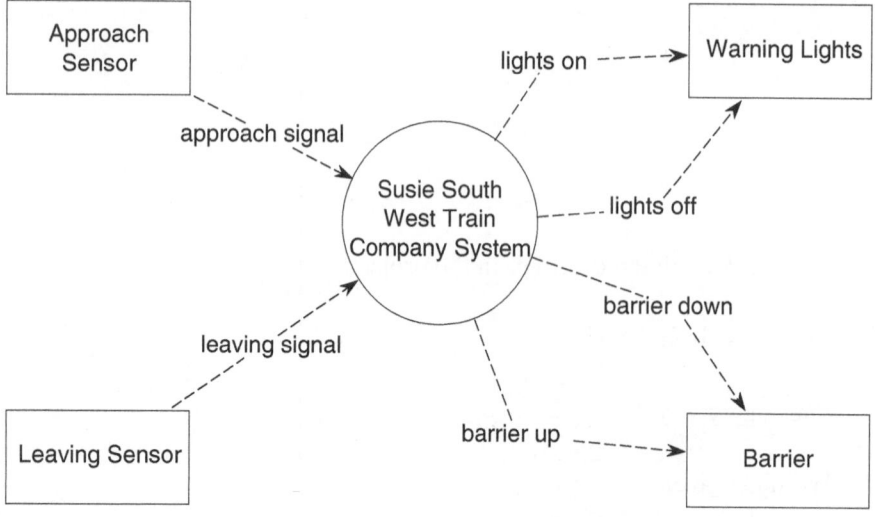

Figure 7.30 *Susie South West Train System: context diagram*

Control refinement

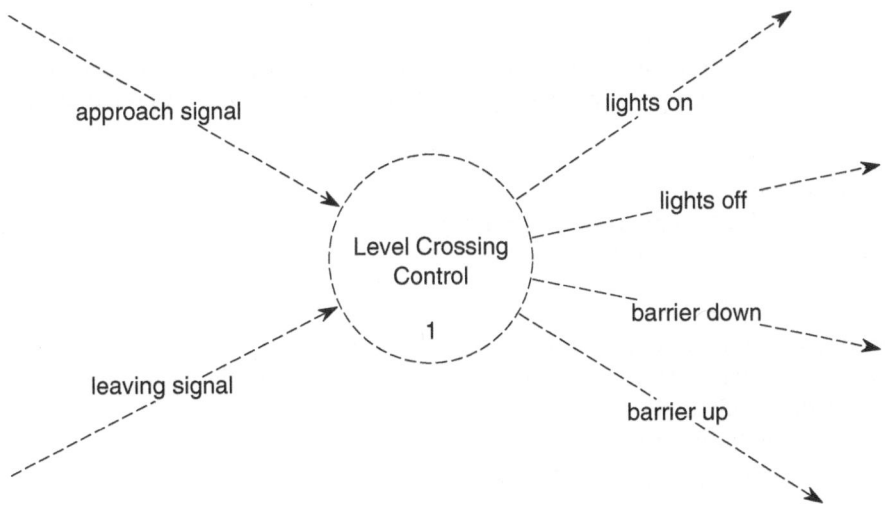

Figure 7.31 *Susie South West Train Company System: control refinement*

State transition diagram

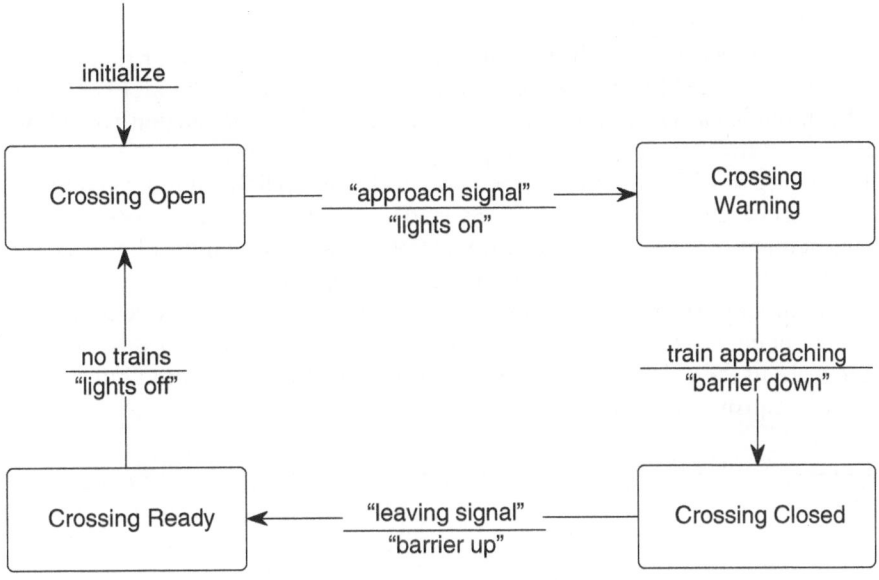

Figure 7.32

Consistency check

Project: C:\SELECT\SYSTEM\TRAIN1\
Title: Susie South West Train Company
Date: 18-Apr-2001 Time: 19:1

Report: Diagram Consistency checking

This report contains a consistency check of all the diagrams in the project.

Checking STRAIN1.DAT
    No Errors detected, No Warnings given.

Checking TRAIN1.DAT
    No Errors detected, No Warnings given.

Checking TRAIN2.DAT
    No Errors detected, No Warnings given.

---- End of report ----

Implementation modelling – some important considerations

To improve the quality of a product a good software designer needs to follow the analysis process with sound design principles. An implementation model represents the concrete features of a system as opposed to an abstract one. The developer will need to consider the structure in detail and base the resulting model on the organizational requirements for the end product. A good implementation model will be based on the following three phases:

- Model the physical processes involved
- Model the software environment in which the software system is to exist
- Model the structure of the software to be produced.

The first two phases may involve modifications to the analysis documentation. Data flow and other development diagrams may need to be modified to take into account the requirements of the system in its operational environment. For example, it may be necessary to link the new system with the company's central computer in order to obtain essential client data.

The third phase translates the implementation-oriented documentation into a system design process using structure charts to represent required program modules. For an object-oriented system the design document will detail the classes required, their variables and method headings. In order to ensure the project is being developed correctly to meet its original aims a thorough verification and validation process must be adopted. This will ensure that the original functionality is maintained and no new functionality added.

Coupling and cohesion

In order to produce a structure chart the developer needs to split the software into component modules. Such modules can be implemented as C functions, BASIC subroutines or Pascal procedures (or functions). The control of these modules needs to be carefully considered and the complexity of the model kept to a minimum. This is best achieved by designing a system where the main section of the program controls a few modules only and other modules are controlled by them (see Figure 7.33).

Coupling is the measure of the degree in which modules depend on each other. High coupling generally implies high complexity and reduced reliability. This can eventually lead to high maintenance costs and difficulties in tracing and correcting semantic or logical errors. Low coupling generally implies lower complexity, improved reliability, lower maintenance costs and easier debugging activities.

Figure 7.33 shows a main process directly controlling six modules. It needs to know the precise interface requirements, the actual parameters being passed and what modules are being accessed. This design does not allow for each module to know about the existence of the other modules or their associated interfaces.

Figure 7.34 outlines a system with improved coupling.

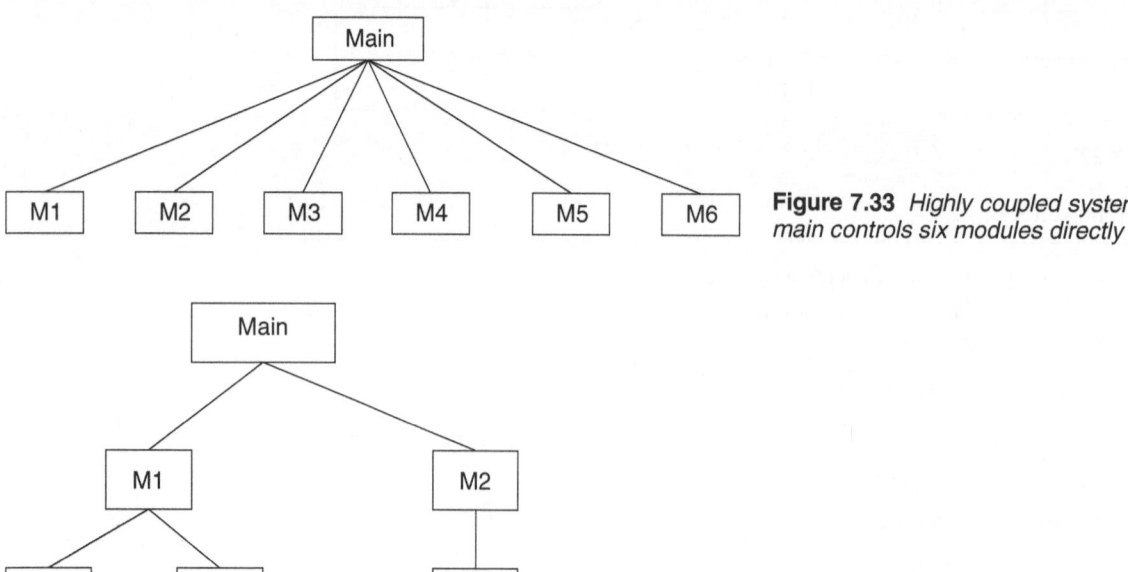

Figure 7.33 *Highly coupled system – main controls six modules directly*

Figure 7.34 *Improved coupling*

As the main section does not have to control as many modules its complexity is reduced. Although modules M1 and M2 have a more complex role as they now have control of subordinate modules. But this approach of reducing complexity from within modules has proved to provide more reliable software with lower maintenance overheads.

Cohesion

This is the measure of how closely the parts of a module correspond to a single function of the software. If any given module is totally concerned with achieving exactly one aim, then cohesion is said to be high. A module is said to exhibit low cohesion if there is not much of a relationship between the elements inside. For example, if a module is designed to ask the user for their name, gets the current date, displays a message and prints an account balance then the module is said to show low cohesion as each of the module aims is not related.

Therefore a good design should show: LOW COUPLING and HIGH COHESION.

Structure chart additional example

College Admission System

Figure 7.35 shows a data flow diagram that represents a College Admission System.

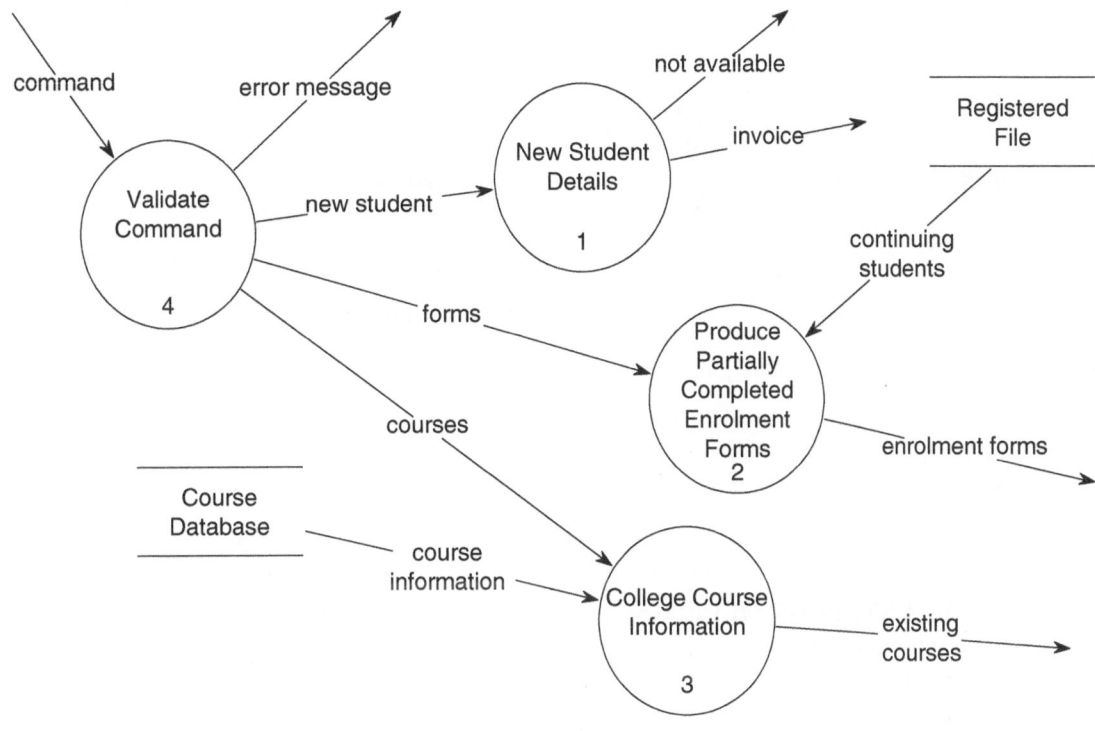

Figure 7.35 *College Admission System*

Complete the following activities:

1. Determine the afferent (input), central transformation, and efferent (output) processes from the diagram above
2. Show how the data flows are to be intersected for the input, central and output requirements suggested in 1. Use Figure 7.35 to show these
3. Complete a structure chart using an appropriate CASE tool to represent the above system
4. Check the diagram is syntactically correct.

Suggested solution – College Admission System

1. Position of processes

Afferent (input) processes	Central transform	Efferent (output) processes
Validate Command	New Student Details Produce Partially Completed Enrolment Forms College Course Information	

2. The flows are split as shown in Figure 7.36

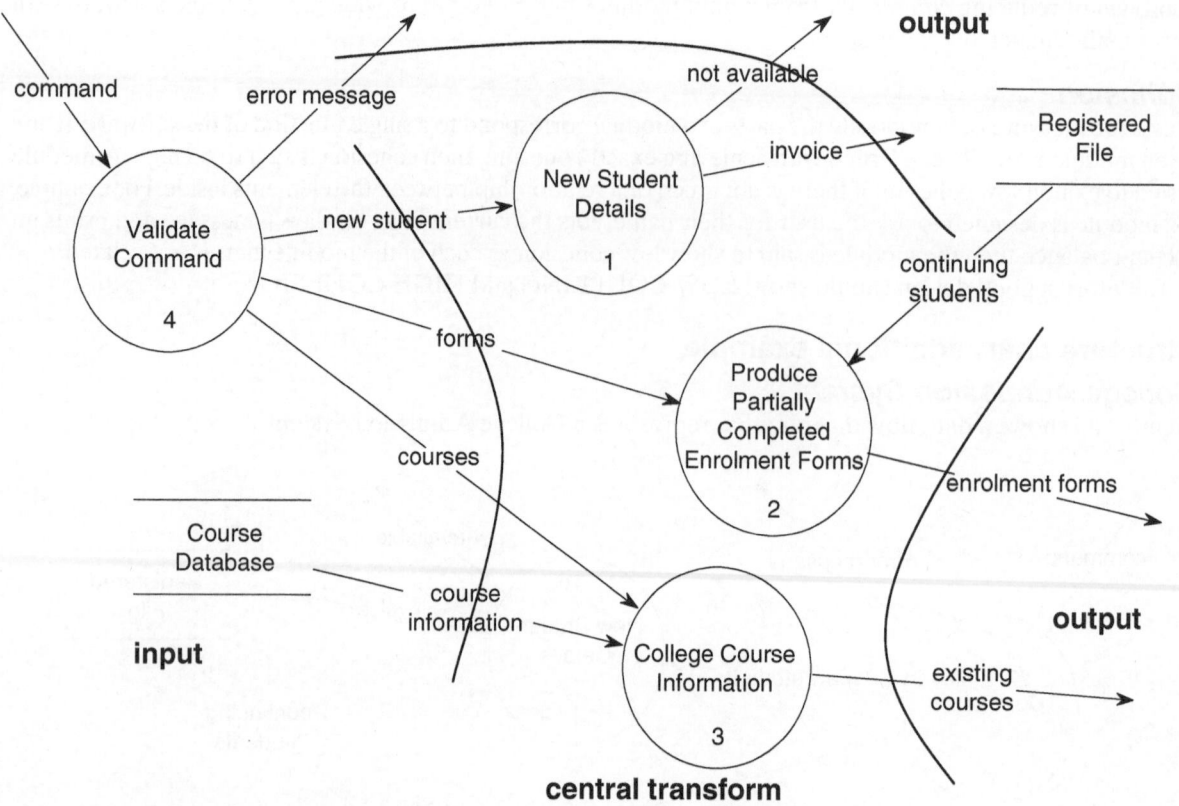

Figure 7.36 *College Admission System*

3. Structure chart. See Figure 7.37 for the structure chart
4. Syntax test

Project: C:\SELECT\SYSTEM\COLLEGE\
Title: College Admission System
Date: 19-Apr-2001 Time: 13:2

Checking CADMIN1.DAT

No Errors detected, No Warnings given.

---- End of report ----

Object-oriented analysis and design

It has been proved that the quality of a software product is vastly improved with the use of object-oriented techniques. This is mainly due to the way object concepts lead themselves to real-world entities and the construction of the class library where developed code can be reused time and time again within given applications.

'Reuse' concerns using or adapting existing software components which are initially provided by the class library supplied with the object system and supplemented by the components constructed by earlier software developments.

Figure 7.38 shows an example from the Java class library showing the Button class.

For the development of Java classes the Java 2 (version 1.3) platform is an essential tool for code development. Note that the top class of the hierarchy is Object and all classes below this can inherit its internal components.

Determining the classes from a given specification was developed in Chapter 2 using textual analysis. This also showed that class structures were the fundamental building block of object development. This is outlined in Figure 7.39.

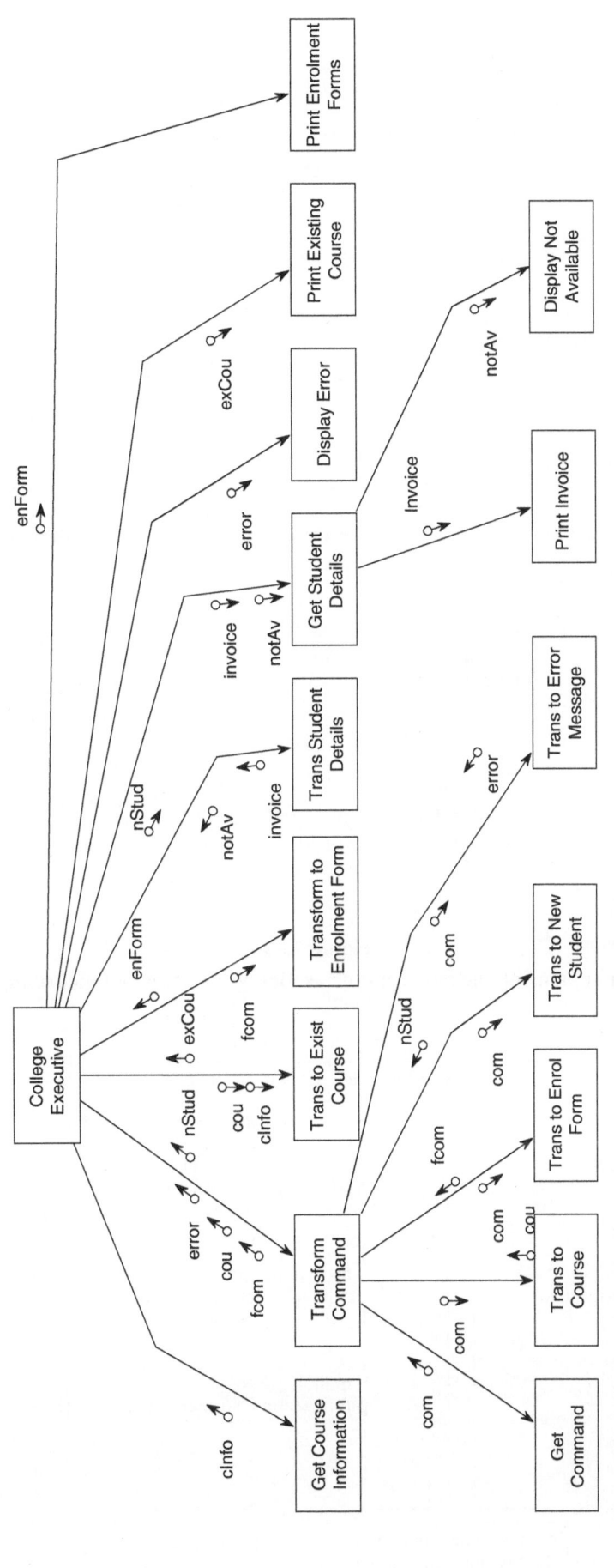

Figure 7.37 Structure chart representation – College Admission System

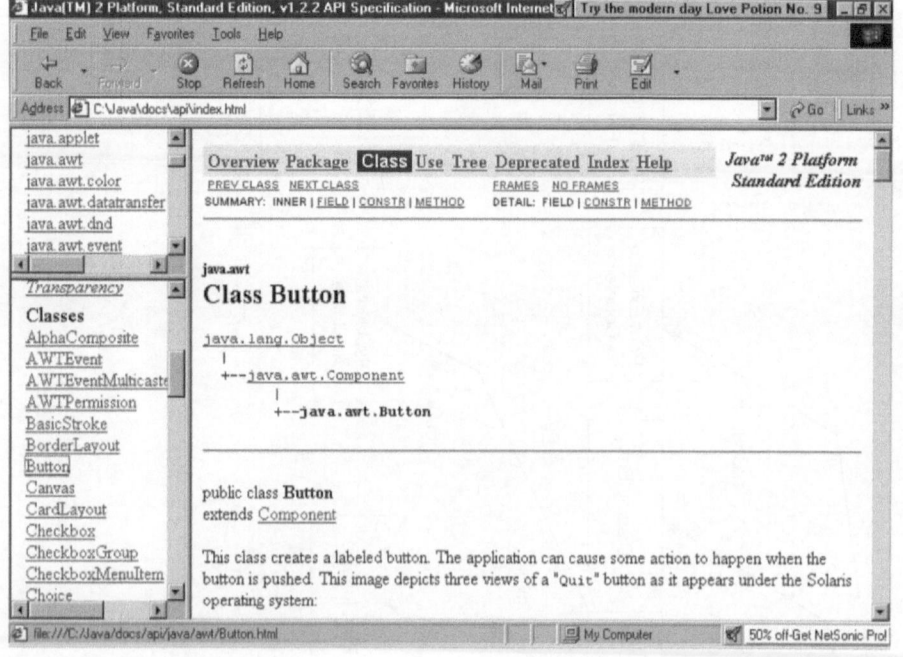

Figure 7.38

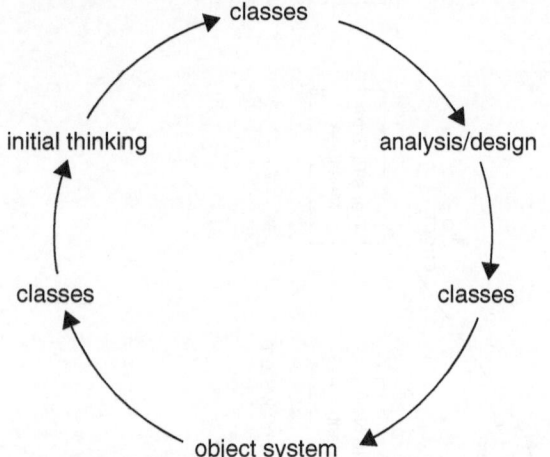

Figure 7.39

Remember a class is an implementable base for objects. Below are some outlines of high level examples of class structures:

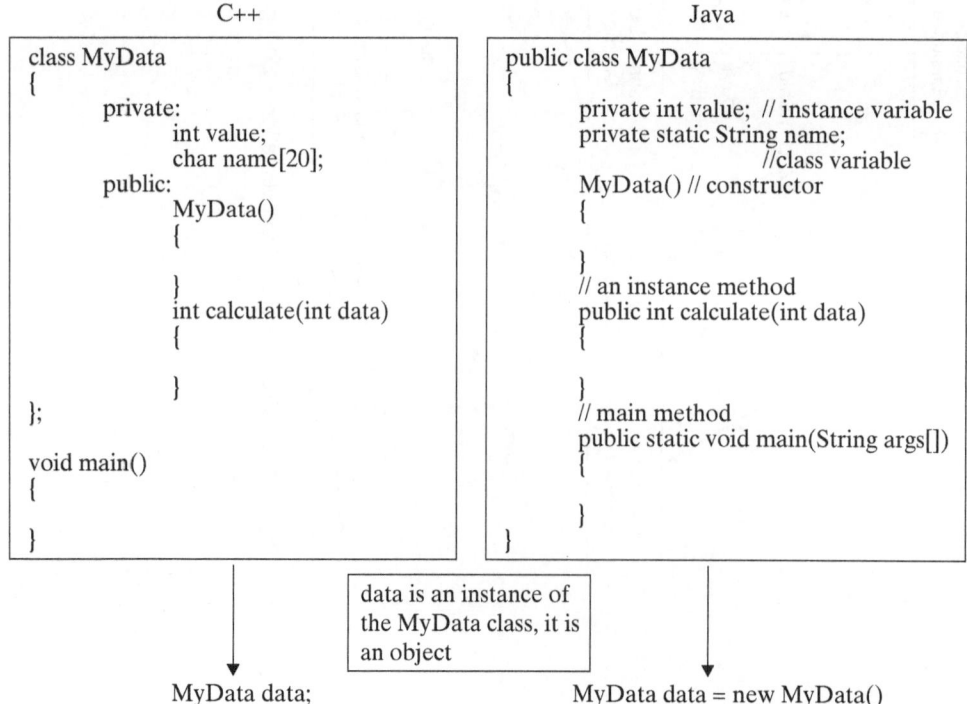

To sum up:

- A class describes the behaviour of a set of objects of some kind
- Objects (which are instances of some class) are related to other objects (mostly, but not necessarily, instances of other classes). The relationships between objects are represented as associations between the classes of which the objects are instances
- The behaviour of a class of objects can be described in terms of the responsibilities that are assigned to the class. There are two forms of responsibilities, for recording information and for doing something (i.e. taking action)
- Objects fulfil their responsibilities by means of collaborations with other objects.

Hence the Class–Responsibility–Collaboration (CRC) method used as a development method within object-oriented software development.

Full details of this method can be found in the following text: *Designing Object-Oriented Software* by Rebecca Wirfs-Brock, Brian Wilkerson and Lauren Wiener, published by Prentice-Hall. The processes involved in this method are outlined in the table below:

Development stage	Activity
Analysis	This initial stage takes a statement of requirements that has been negotiated with the customer and is analysed to obtain an understanding as to what the system is to do. The aim is to provide an initial understanding of the classes, associations, responsibilities and collaboration. Here the initial class-association diagram will be developed possibly using techniques like UML with an appropriate CASE tool which provide checking facilities to ensure quality is built into the model from the start.
Design	This stage is concerned with producing a design document that specifies how the object model may be represented by software components. The design document will detail the classes required, their method heading and the associated variables. This stage moves away from real-world entities and events to objects as software components. An example table format outlining the CRC, method and variable class requirements is shown below.
Implementation	This stage takes the design activity and essentially produces the appropriate object code. For example C++, Smalltalk or Java.
Component and integration testing	The individual components are tested first to ensure their internal protocol is correct. Then the components are brought together to form the complete working system. During this stage integration testing is carried out to ensure the components correctly link together and all message calls between classes are correctly structured.

Class design table:

Class: MyData			Superclass: SuperData
Responsibility	Type	Collaborator	Variables and methods
Records the data details	record	YourData	variables:
Calculate the data fee	action	None	methods:

Note: The number of responsibilities contained within a class should be kept to a minimum. An average figure is normally around two, with four being the maximum.

At the design stage we need to think carefully about the class protocol (the methods) and their associated variables. Most modern object-oriented languages classify their variables and methods as being 'class' or 'instance'.

The piece of Java code shown previously contains examples of these. Another example is shown below:

```
private static char letter;              // class variable
public static int myMethod(String name)  // class method
{
        ....................;
        ....................;
        return ( );

}
```

These are specified as being a 'class variable' and a 'class method' . This is defined within Java by the use of the static term.

If we have:

private double balance;   // instance variable

public void setData(String data)   // instance method
{

}

So what are the differences?

Instance variable: These hold unique values for each instance of the class created. For example, if we have a class called Account, which has an instance variable called balance, and we create two instances of the class (objects), baileyAccount and bessAccount, then each can have a unique balance value.

Class: Account

Instances of the class (objects)	Instance variable	Value
baileyAccount	balance	579.23
bessAccount	balance	23.57

Generally instance variables can only be directly accessed from the methods of the class in which they are defined and from the methods of its subclasses.

Instance method: A method is the implementation of the message that has the same name. A method is the code that is executed when the corresponding message is sent to an object. These may be used as 'accessor methods' for the instance variables. For example:

accessor methods

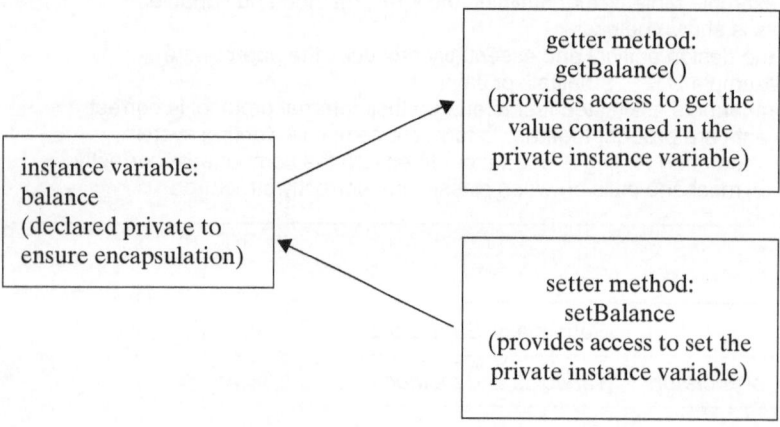

Note: An instance method is sent to an instance of a class (i.e. an object), i.e.

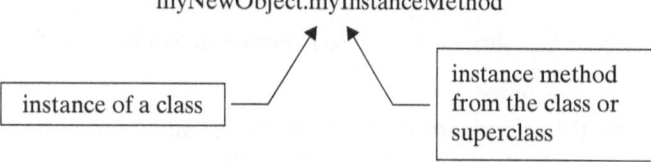

Class variable: This is the most common means by which classes can have state associated with them. These variables are visible to and can be used directly by the class and instance methods of the class, its subclasses, instances of the class and instances of any subclasses. Hence they allow the class state to be shared by all instances, subclasses, and their instances. A class variable is similar to a global variable in that it all instances of the class share its same common state.

Class method: Class methods can be used in a similar way as instance methods for providing 'assessor routines' for the class variables. They can be used without creating an instance of a class. A class method is defined for a message sent directly to a class. For example:

MyClass.myClassMethod

Note: Instance methods can also access class variables, but class methods cannot directly access instance variables. Instance methods can also access class methods from within their code structure.

Object-oriented example – airline booking system

Below is an extract from a negotiated statement of requirements:

An airline booking system is required that can take passenger reservations for specific flights. The system needs to keep track of many airlines each with its own respective flights. It will handle all bookings and cancellations made by customers it has, or will service. Each flight will contain many passengers that are made up of first class, business class and tourist class. Also each flight will need to keep track of its flight number and destination plans.

1. Analyse the statement of requirements to establish the required classes
2. Using a suitable CASE tool construct an outline class-association diagram and check its syntactic correctness
3. Develop the class text structures that represent the diagram outlines in 2 above
4. Consider any invariants that may need to be documented
5. Carry out a 'walkthrough' to ascertain the responsibilities and collaborations
6. Develop any aspects of inheritance that may be explicit in the analysis model
7. Complete a design table to represent the classes to include their variables and methods. The type of responsibility needs to be included along with the collaborators established during the 'walkthrough'
8. Modify the class diagram to include the variables and methods (operations)
9. Complete a basic Java object-oriented programming implementation of the analysis and design model to contain part of the AirlineBookingSystem that is able to create two new airline objects and one tourist class passenger object. You will need to complete four classes: AirlineBookingSystem, Airline, Passenger (abstract) and TouristClassPassenger (which inherits from Passenger). Remember to create a new tourist class passenger object you need the following code structure:
TouristClassPassenger newPassenger = new TouristClassPassenger();
10. Thoroughly test the classes by adding a test harness (main program section) in the AirlineBooking System class, and finally document the execution results.

Note: Remember the CRC (class, responsibility and collaboration) method provides a basis for us to get started. The problem specified above would be a massive coding project, but we are just looking at the structures contained in the extract (i.e. showing collaboration and inheritance). Do not go beyond the specification as this may lead to an overcomplicated solution.

For implementation the responsibilities and collaborators will enable you to determine the variable structures and their supporting methods. Each class will require a constructor that will set the initial values of the instance variables.

As this is not an object-oriented programming course (full details in the unit of the same name) keep the storage structures simple. For example, use array structures and not complicated data storage techniques. Remember that Passenger is an abstract class in which no instances need to be created; its subclasses (TouristClassPassenger, FirstClassPassenger and BusinessClassPassenger) all inherit from the superclass.

Suggested solution – airline booking system

1. By using the process of textual analysis we have:

An <u>airline booking system</u> is required that can take <u>passenger reservations</u> for specific <u>flights</u>. The system needs to keep track of many <u>airlines</u> each with its own respective <u>flights</u>. It will handle all bookings and cancellations made by <u>customers</u> it has, or will service. Each <u>flight</u> will contain many <u>passengers</u> that are made up of <u>first class</u>, <u>business class</u> and <u>tourist class</u>. Also each <u>flight</u> will need to keep track of its flight number and destination plans.

The main candidates for class structures are:

Airline Booking System (AirlineBookingSystem)
Airline
Flight
Passenger
    Passenger First Class (FirstClassPassenger)
    Passenger Business Class (BusinessClassPassenger)
    Passenger Tourist Class (TouristClassPassenger)

Note: Customers are future passengers so a class for them is not required.

2. Outline class structure

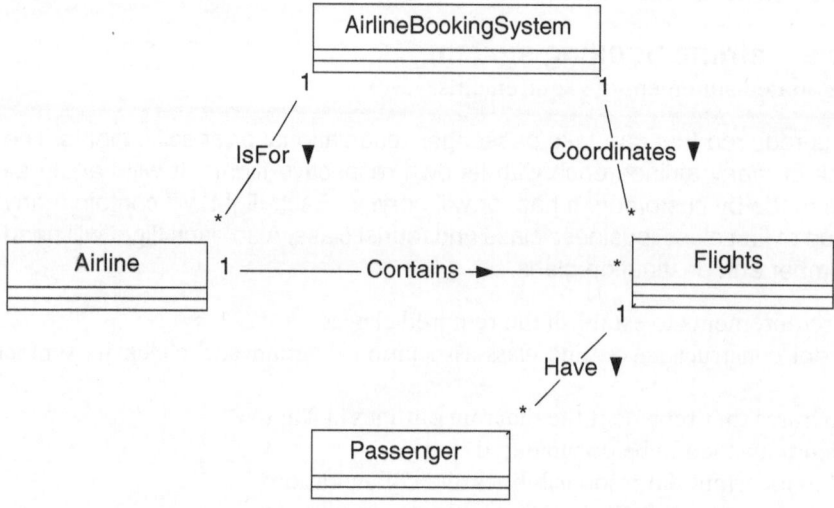

Figure 7.40

Note: This diagram was completed using Select Enterprise© CASE tool.

To ensure quality is installed from the outset the diagram was checked for semantic correctness. At the moment it is an outline only and may require modifications as the model develops.

3. Develop the class text structures that represent Figure 7.40 above
Class: AirlineBookingSystem
'Handles all the booking requirements for the particular airlines and passengers'

Class: Airline
'Provides overall control for all its flights'

Class: Flight
'Provides details about particular flights'

Class: Passenger
'Passengers travel or make booking to travel'

Class: FirstClassPassenger
'A type of passenger that only travels first class'

Class: BusinessClassPassenger
'A special type of passenger that only travels business class'

Class: TouristClassPassenger
'A passenger who travels by the cheapest rate'

4. Consider any invariants that may need to be documented
Invariant
Any given instance of Flight is associated with a set of Passenger instances via the contains association.
Each instance of passenger is either: First Class, Business Class or Tourist Class.

5. Carry out a 'walkthrough' to ascertain the responsibilities and collaborations
Walkthrough
Stage 1: Understanding in application area terms
What information is the system provided with?
1. Passenger details (name, address, travel requirements etc.)
2. The Airline and Flight details
What must the system do with that information?
1. Record the passenger information
2. Record that a passenger is booked onto a flight associated to a particular airline.

Stage 2: Understanding in terms of instances of classes and associations
Processing the booking of a passenger for a flight requires the following actions:
1. Locating the instance of Flight corresponding to the scheduled flight

2. Navigating via the contains association to those instances of Passenger that correspond to those passengers belonging to that flight

3. For each such instance of Passenger, obtain the value corresponding to the passenger's name.

Stage 3: Understanding in terms of responsibilities and collaborations
Class: AirlineBookingSystem
Responsibilities:
   Keeps track of the airlines under its control
   Keeps track of the passengers it has or will service
   Handles passenger bookings
Collaborators:
   Airline, Passenger

Class: Airline
Responsibilities:
   Records the airline name
   Keeps track of the scheduled flights
Collaborators:
   Flight

Class: Flight
Responsibilities:
   Records the flight name
   Keeps track of the passengers in a flight
   Records the route and flight times
Collaborators:
   Passenger, Airline

Class: Passenger
Responsibilities:
   Records the passenger name, address and telephone number
Collaborators:
   None

Class: FirstClassPassenger
   'A subclass of passenger'
Responsibilities:
   Records the passenger's special requirements
Collaborators:
   None

Class: BusinessClassPassenger
   'A subclass of passenger'
Responsibilities:
   Records the passenger's company name
Collaborators:
   None

Class: TouristClassPassenger
   'A subclass of passenger'
Responsibilities:
   Records the passenger's travel company
Collaborators:
   None

6. Develop any aspects of inheritance that may be explicit in the analysis model

From the class structure above we can see that Passenger is a prime candidate for a superclass of Business Class Passenger, First Class Passenger and Tourist Class Passenger. As passengers can only be of one type only Passenger can be defined as abstract as no instance of it need to be created.

All the three subclasses can inherit the methods created in Passenger as they are common to all of them. The layout is shown in Figure 7.41.

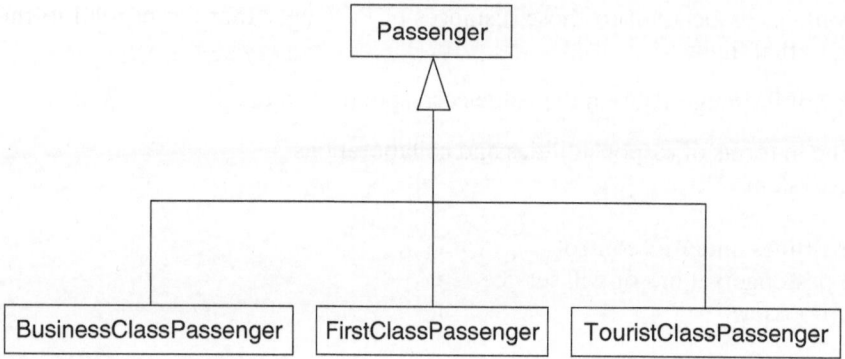

Figure 7.41

7. Complete a design table to represent the classes to include their variables and methods. The type of responsibility needs to be included along with the collaborators established during the 'walkthrough'

Class: AirlineBookingSystem			Superclass: Object
Responsibilities	Type	Collaborators	Variables and methods
Keeps track of the airlines under its control	record	Airline	Instance variables: airlines passengers
Keeps track of the passengers it has or will service	record	Passenger	Instance methods: addAirlines() addPassengers()
Handles passenger bookings	record	Passenger	Instance methods: bookPassenger()

Class: Airline			Superclass: Object
Responsibilities	Type	Collaborators	Variables and methods
Records the airline name	record		Instance variables: airlineName flights
Keeps track of the scheduled flights	record	Flights	Instance methods: setAirline() getAirline()

Class: Flight			Superclass: Object
Responsibilities	Type	Collaborators	Variables and methods
Records the flight name	record		Instance variables: flightName flightPassengers
Keeps track of the passengers in a flight	record	Passenger	route flightTimes
Records the route and flight times	record		Instance methods: setFlightName() getFlightName() addPassengers() getPassengers() setRoute() getRoute() setFlightTimes() getFlightTimes()

Class: Passenger			Superclass: Object
Responsibilities	Type	Collaborators	Variables and methods

Records the passenger name, address and telephone number	record	none	Instant variables
			   name
			   address
			   phone
			Instance methods
			   getName(), setname()
			   getAddress(), setAddress()
			   getPhone(), setPhone()

Class: BusinessClassPassenger			Superclass: Passenger
Responsibilities	Type	Collaborators	Variables and methods

Records the passenger's company name	record	none	Instance variables:
			   companyName
			Instant methods:
			   setCompanyName()
			   getCompanyName()

Class: FirstClassPassenger			Superclass: Passenger
Responsibilities	Type	Collaborators	Variables and methods

Records the passenger's special requirements	record	none	Instance variables:
			   specialRequirements
			Instant methods:
			   setSpecialRequirements()
			   getSpecialRequirements()

Class: TouristClassPassenger			Superclass: Passenger
Responsibilities	Type	Collaborators	Variables and methods

Records the passenger's travel company	record	none	Instance variables:
			   travelCompany
			Instant methods:
			   setTravelCompany()
			   getTravelCompany()

8. Modify the class diagram (developed in 2) to include the variables (attributes) and methods (operations)

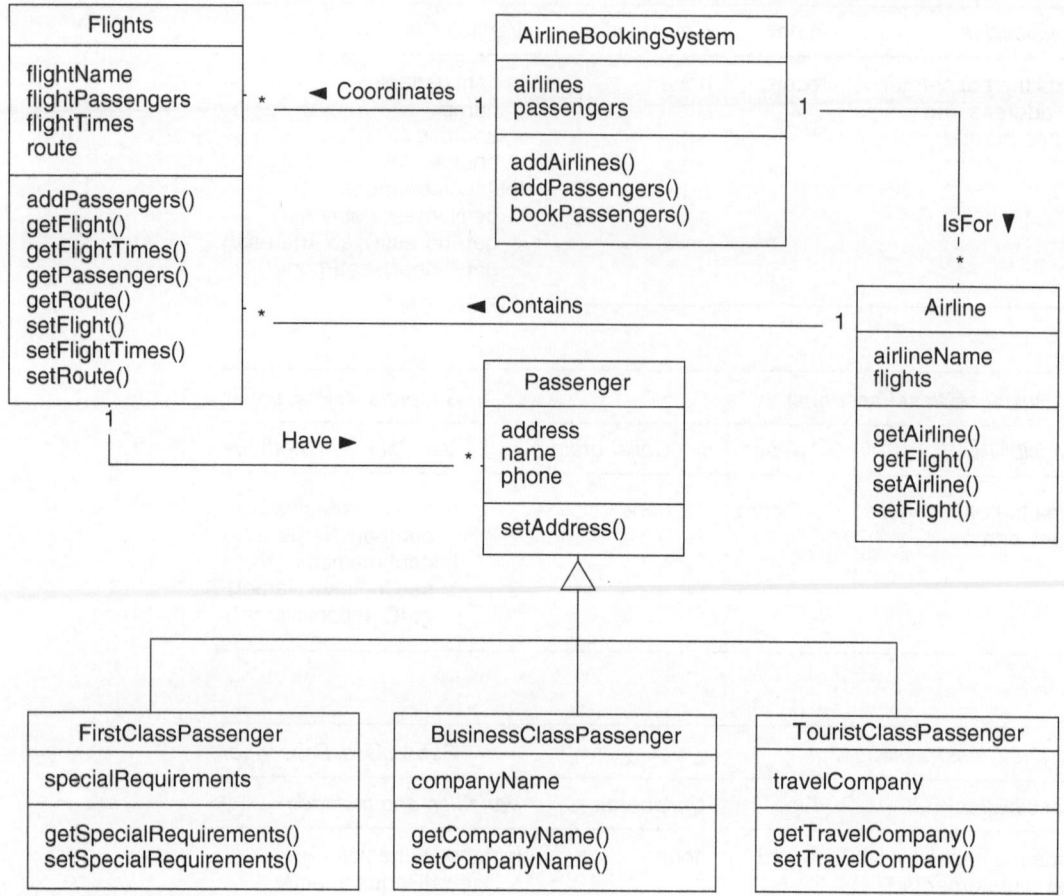

Figure 7.42 *Syntax check*

```
Project: N:\MYWORK\SELECT\SYSTEM\AIR1\
Title: Airline Booking System
Date: 7-Jun-01 Time: 11:40

Checking AIR1.DAT

No Errors detected, No Warnings given.

---- End of report ----
```

9. Complete a Java object-oriented programming implementation of the analysis and design model developed so far
   Outline Class Structures:

```java
/* Airline Booking System main class
AirlineBooking System.java
*/

public class AirlineBookingSystem
{
        // instance variables
        private String airlines[];
        private String passengers[];
        // class variables - required for input control
        static byte buffer[];
        static String inString, trimString;

        AirlineBookingSystem()     // constructor
        {
                airlines = new String[10];
                passengers = new String[10];
        }
```

```
public void addAirlines()
throws java.io.IOException
{
        String anAirline;
        //create two airline objects by collaborating with Airline
        Airline newAirline1 = new Airline();
        Airline newAirline2 = new Airline();
        //enter first airline name
        System.out.println("\nAdd new airlines: \n");
        System.out.print("Create first new airline name: ");
        buffer = new byte[80];
        System.in.read(buffer);
        inString = new String(buffer);
        trimString = inString.trim();
        airlines[1] = trimString;
        newAirline1.setAirlineName(airlines[1]);
        //enter second airline name
        System.out.print("Create second new airline name: ");
        buffer = new byte[80];
        System.in.read(buffer);
        inString = new String(buffer);
        trimString = inString.trim();
        airlines[2] = trimString;
        newAirline2.setAirlineName(airlines[2]);
        // test objects have been successfully created
        System.out.println("\n\nThe following airlines have been
        created:\n"+newAirline1.getAirlineName() + " and " +
        newAirline2.getAirlineName());
}

public void addPassengers()
throws java.io.IOException
{
        //create new tourist class passenger object (newPassenger)
        //note the collaboration with the Passenger class
        TouristClassPassenger newPassenger = new
        TouristClassPassenger();
        String newPhone, newAddress, newTravelCompany;
        System.out.println("\nAdd new passenger details: \n");
        //enter the passengers name and store in an array for
        //future use
        System.out.print("Enter the passengers name: ");
        buffer = new byte[80];
        System.in.read(buffer);
        inString = new String(buffer);
        trimString = inString.trim();
        passengers[1] = trimString;
        newPassenger.setName(passengers[1]) ;
        //enter the passengers address
        System.out.print("\nEnter the passengers address: ");
        buffer = new byte[80];
        System.in.read(buffer);
        inString = new String(buffer);
        trimString = inString.trim();
        newAddress = trimString;
        newPassenger.setAddress(newAddress);
        //enter the passengers phone number
        System.out.print("\nEnter the passengers phone number: ");
        buffer = new byte[80];
        System.in.read(buffer);
        inString = new String(buffer);
        trimString = inString.trim();
        newPhone = trimString;
        newPassenger.setPhone(newPhone);
        //enter the travel company
        System.out.print("\nEnter the passengers travel company: ");
        buffer = new byte[80];
        System.in.read(buffer);
        inString = new String(buffer);
        trimString = inString.trim();
        newTravelCompany = trimString;
        newPassenger.setTravelCompany(newTravelCompany);
```

```
                    //test the new passenger has been created
                    System.out.println("\nCheck of new passenger object: ");
                    System.out.println("\nPassenger Name: " + newPassenger.getName());
                    System.out.println("\nPassenger Address: " + newPassenger.getAddress());
                    System.out.println("\nPassenger Phone: " + newPassenger.getPhone());
                    System.out.println("\nPassenger Travel Company: "
                                        + newPassenger.getTravelCompany());
            }

        //test harness to check the above methods
        public static void main(String args[])
        throws java.io.IOException
        {
            int check;
            AirlineBookingSystem aSystem = new AirlineBookingSystem();
            do
            {
                System.out.print
                    ("\nSelect 1 for Airlines or 2 for Passengers (3 to
                    end): ");
                buffer = new byte[80];
                System.in.read(buffer);
                inString = new String(buffer);
                trimString = inString.trim();
                check = Integer.parseInt(trimString);
                if ( check == 1 )
                {
                    aSystem.addAirlines();
                }
                else
                    if ( check == 2 )
                    {
                        aSystem.addPassengers();
                    }
            }
            while ( check ! = 3 );
            System.out.println("\n\nEnd of Program .................. ");
        } // end main

}
```

Airline class:
```
/* Airline class for Booking System
   Airline.java
*/

public class Airline
{
        // instance variables
        String airlineName, flights;

        Airline()        //constructor
        {
                airlineName = null;
                flights = null;
        }


        //basic instance methods for the airline name i.e. Canada Air
        public void setAirlineName(String aName)
        {
                airlineName = aName;
        }

        public String getAirlineName()
        {
                return airlineName;
        }
}
```

Passenger class:

```
/* Passenger class to maintain customer that is entered
   via the AirlineBookingSystem class
   Passenger.java
*/
abstract class Passenger
{
        // instance variables
        private String name;
        private String address;
        private String phone;
        Passenger()    //constructor
        {
                name = "name";
                address = "empty address";
                phone = "no phone";
        }
        // instance methods
        public String getName()
        {
                return name;
        }
        public void setName(String aName)
        {
                name = aName;
        }
        public String getAddress()
        {
                return address;
        }
        public void setAddress(String anAddress)
        {
                address = anAddress;
        }
        public String getPhone()
        {
                return phone;
        }
        public void setPhone(String aPhone)
        {
                phone = aPhone;
        }
}
```

Tourist class passenger:

```
/* This is a sub-class of Passenger for
   the tourist class passengers.
   TouristClassPassenger.java
*/
public class TouristClassPassenger extends Passenger
{
        // instant variable
        private String travelCompany;
        TouristClassPassenger() //constructor
        {
                travelCompany = "Travel Company";
        }
        // instance methods
        public void setTravelCompany(String aCompany)
        {
                travelCompany = aCompany;
        }
        public String getTravelCompany()
        {
                return(travelCompany);
        }
}
```

10. Thoroughly test the classes by adding a test harness (main program section) in the AirlineBooking System class, and finally document the execution results

```
C:\Java\bin>java AirlineBookingSystem

Select 1 for Airlines or 2 for Passengers (3 to end): 1

Add new airlines:

Create first new airline name: Canada Air
Create second new airline name: West Pacific

The following airlines have been created:
Canada Air and West Pacific
```

```
Select 1 for Airlines or 2 for Passengers (3 to end): 2

Add new passenger details:

Enter the passenger's name: Fred Smith

Enter the passenger's address: 22 West Side Lane,
Surbiton, Surrey

Enter the passenger's phone number: 020 8766 6565

Enter the passenger's travel company: Sunshine Tours

Check of new passenger object:

Passenger Name: Fred Smith

Passenger Address: 22 West Side Lane, Surbiton, Surrey

Passenger Phone: 020 8766 6565

Passenger Travel Company: Sunshine Tours
```

Software metrics

Further information

We have seen that software metrics provide numerical measures of the products and processes that make up the software product. Traditionally metrics have been applied at the later stages of a software project, but this limits their usefulness in predicting the course it takes and monitoring the controlling processes. For metrics to be fully useful as a quality assurance tool they should be used throughout the lifecycle of the development process. Two examples of metrics that may be applied at the early stages of development are outlined below:

- Function-point analysis which concentrates on extracting information from the system specification to measure productivity and to predict product cost
- Maintenance analysis which concentrates on extracting information from the system's architecture, produced during system design to minimize the cost of system design.

A selection of where to possibly obtain measurable metrics is outlined below:

- The number of decisions in the detailed design of a program unit
- The length of a program unit or object class
- The number of discovered errors in a software system during the first 12 months of operation
- The amount of effort put into a project
- The number of classes in an object-oriented design diagram for a software system.

These are not all useful, and the following three tests should be applied to any potential metric:

- The purpose of a metric should be clear and concise
- The model that the metric is to be used on should be tested for suitability
- When applied the metric should be analysed to ensure it is the most suitable method.

Some main characteristics of a software metric:

- A software metric must be measurable
- A software metric must be independent
- A software metric must be accountable
- A software metric must be precise.

A software metric should be based on facts not opinions, it should not be possible for project team members to alter its value without effecting the quality of the software. Details of how the metric was measured should be documented and the level of tolerance allowed when measuring it should be specified.

Examples of metrics

- *Size metrics* – Relates to the size of the software code
- *Data structure metrics* – This analyses data in the form:

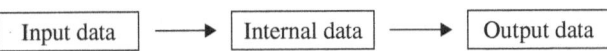

- *Logic structure metrics* – How the program allows different operations dependent upon different input data or intermediate calculations
- *Composite metrics* – This looks at both size and logic structure metrics to form an order hierarchy based on complexity
- *Effort and cost metrics* – This can be used to model the project by bringing in the proposed costs and estimates that may occur. It may also be used to outline errors in calculations by comparing 'Results metrics' against 'Predictor metrics', see Figure 7.43.

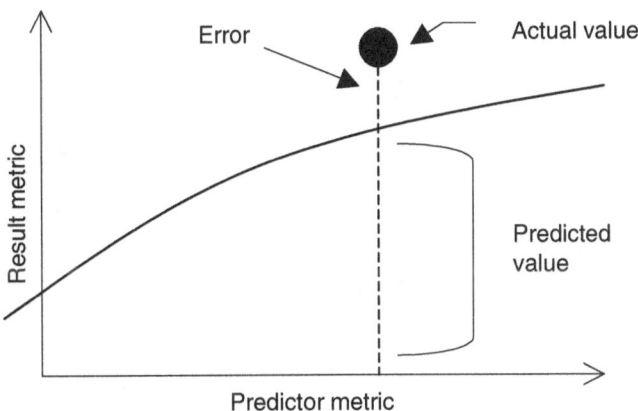

Figure 7.43

Error = Actual value − Predicted value

The historical prediction errors are used to estimate the model.

- *Design metrics* – In general program complexity increases with program size. Design metrics are used to encourage program modularization, i.e. ensure a structured approach by partitioning work at an early stage. Design metrics can be used to analyse structure charts and data flow diagrams (DFDs).

McCabe's Cyclomatic Complexity metrics
Further example
Consider the following program fragment:

```
int total = 0;
int index = 1;
int count = 0;
do
{
        leave = false;
        jindex = 1;
        do
```

```
                        {
                                if (temp[index] = ignore[jindex])
                                {
                                        leave = true;
                                }
                                jindex++;
                        }
                while (jindex <= 10);
                cout << "Test Output: "<< index;
                if ( !leave )
                {
                        total += temp[index];
                        count++;
                }
                index++;
        }
while (temp[index] ! = -1);
cout << "Total:"<< total
if ( index > 1 )
{
        average = total/count;
}
cout << "Average: "<< average;
```

Questions

1. Draw a graph to represent the code structure
2. Calculate its Cyclomatic Complexity
3. Comment on the complexity of the design.

Suggested solution – Cyclomatic Complexity example

1. Graph

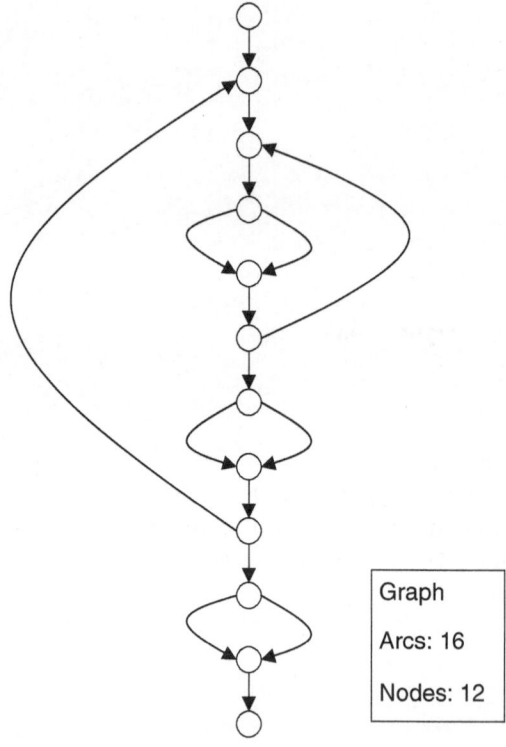

Graph

Arcs: 16

Nodes: 12

Figure 7.44

2. Cyclomatic Complexity:
   Arcs − Nodes + 2
   16 − 12 + 2 = 6
   Cyclomatic Complexity: 6

3. The design is fairly complex but as the cyclomatic complexity is <10 it does not warrant any redesigning. This is based on the research carried out by McCabe and backed up with experimental evidence.

Impact matrices

Impact matrices are used to help determine the amount of resource that a particular system is going to use during the process of maintenance. A project manager wants to know as early as possible in the development process the risk factors that may affect the software system during its lifetime.

An impact matrix is used to display the change (impact) of one unit (or method within an object-oriented class structure) on another. The matrix is a two-dimensional table that shows the probability that a modification of one unit will directly lead to the need to change another unit. The configuration manager possibly using CASE tool analysis usually determines the probabilities.

The main aim at this stage is the interaction and use of the probabilities, not how they were derived at.

Example

	Impact matrix				
	V	W	X	Y	Z
V	1.0	0.3	0.2	0.1	0
W	0.3	1.0	0.4	0	0
X	0.2	0.4	1.0	0	0.1
Y	0.1	0	0	1.0	0.3
Z	0	0	0.1	0.3	1.0

The matrix is assumed to be symmetric. For example, a change in unit 'V' affects unit 'W' by 0.3 (30%) and the reverse is true, a change in unit 'W' affects unit 'V' by the same amount (0.3 or 30%). This may not always be the case in all applications, but for this example we are considering the matrix to be symmetric.

The second row shows that if there is a change to 'W' then there is a probability of 0.3 that, as a direct result of a change in 'W', there will be a change to 'V'. A probability of 0.4 that there will be a change to 'X' and no change to 'Y' or 'Z'.

Note: There must be a probability of 1.0 (100%) of a change to a unit affecting itself.

This can be represented as a block diagram:

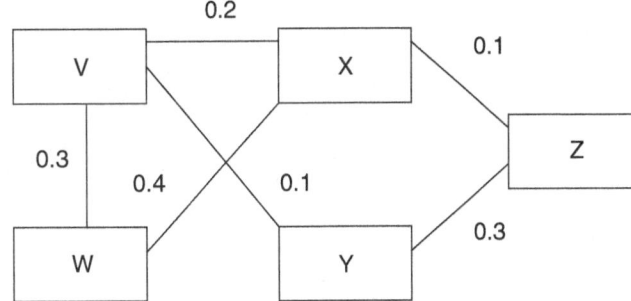

Figure 7.45

From the diagram it is easy to see that the unit 'V' can affect unit 'X' indirectly via unit 'W' as well as directly. The probability of this second order effect from 'V' to 'X' is calculated as follows:

- A change in program unit 'V' has a probability of 0.3 of affecting program unit 'W'
- A change in program unit 'W' has a probability of 0.4 of affecting program unit 'X'
- Hence a change to program unit 'V' will have a probability of affecting program unit 'X' by means of a second order effect of $0.3 \times 0.4 = 0.12$.

Question

Calculate the probability of a change in program unit 'V' affecting program unit 'Z' as a result of second order effects.

Suggested solution

There are two second order effects of 'V' on 'Z':

1. First via 'X' ('V' to 'X' with a probability of 0.2 and 'X' to 'Z' with a probability of 0.1). This gives a total second order effect of 0.02

2. Second via 'Y' ('V' to 'Y' has a probability of 0.1 and 'Y' to 'Z' has a probability of 0.3). This gives a total second order effect of 0.03.

This gives a total second order effect of a change to unit 'V' affecting unit 'Z' of $0.02 + 0.03 = 0.06$.

Note: There are other higher order effects, for example 'V' to 'W' to 'X' then 'Z' and in complex systems these would need to be considered. This process is outlined in the following example.

Let us now consider the overall effect of 'V' on 'X'. From the diagram it can be seen that 'X' is affected directly from 'V' (first order effect) and indirectly through 'W' (second order effect). The probabilities are 0.2 and 0.12 (see above). Units 'V' and 'X' also interact via 'Y' (giving a third order effect) this has the probabilities of 'V' to 'Y' 0.1, 'Y' to 'Z' 0.3 and 'Z' to 'X' 0.1, giving $(0.1 \times 0.3 \times 0.1) = 0.003$. This figure needs to be calculated in the overall value. In order to do this the following formula is used:

$$P(VX) = P(VX_1) + P(VX_n) - P(VX_1)P(VX_n)$$

This gives a total possibility of change from 'V' to 'X' of:

$$
\begin{aligned}
P(VX) &= P(VX_1) + P(VX_n) - P(VX_1)P(VX_n) \\
&= 0.2 + (0.12 + 0.003) - (0.2 \times (0.12 + 0.003)) \\
&= 0.2 + (0.123) - (0.2 \times 0.123) \\
&= 0.323 - 0.0246 \\
&= \mathbf{0.2984} \\
&= \text{Total probability of change between units 'V' and 'X'}
\end{aligned}
$$

The total probability of all possible interactions is finally displayed on a dependency matrix. The steps taken to achieve this are outlined below:

*   Identify any second or higher order effects between units or methods whose dependency is represented on the block diagram
*   Calculate the probability associated with each of the second or higher order effects. They may be more than one.
*   Calculate the total probability of a change in one unit or method on another using the formula outlined above
*   Construct a dependency matrix.

Below is the dependency matrix with the changes in 'V' affecting 'X' added:

Dependency matrix

	V	W	X	Y	Z
V	1.0		**0.2984**		
W		1.0			
X	**0.2984**		1.0		
Y				1.0	
Z					1.0

Note: It is easy to make a mistake with the calculations. The result must be slightly larger than the original impact matrix figure if second or higher level effects are taken into account.

Question
Calculate the effects of a change to 'V' affecting 'Y' and add the results to the dependency matrix.

Suggested solution
There are two higher level routes from 'V' to 'Y'

1. $V - X - Z - Y = (0.2 \times 0.1 \times 0.3) = 0.006$
2. $V - W - X - Z - Y = (0.3 \times 0.4 \times 0.1 \times 0.3) = 0.0036$

$$
\begin{aligned}
\text{Therefore:} \quad P(VY) &= 0.1 + (0.006 + 0.0036) - (0.1 \times (0.006 + 0.0036)) \\
&= 0.1 + 0.0096 - (0.1 \times 0.0096) \\
&= 0.1096 - 0.00096 \\
&= \mathbf{0.10864}
\end{aligned}
$$

			Dependency matrix		
	V	W	X	Y	Z
V	1.0		0.2984	**0.10864**	
W		1.0			
X	0.2984		1.0		
Y	**0.10864**			1.0	
Z					1.0

The rest of the dependencies can then be calculated to complete the table.

Appendix A
Fact sheet

ASCII character set presented as control characters and text characters

The ASCII control characters, values 0–31

Dec	Hex	Keyboard	Binary	Description	
0	0	CTRL @	00000	NUL	Null Character
1	1	CTRL A	00001	SOH	Start of Heading
2	2	CTRL B	00010	STX	Start of Text
3	3	CTRL C	00011	ETX	End of Text
4	4	CTRL D	00100	EOT	End of Transmission
5	5	CTRL E	00101	ENQ	Enquiry
6	6	CTRL F	00110	ACK	Acknowledge
7	7	CTRL G	00111	BEL	Bell or beep
8	8	CTRL H	01000	BS	Back Space
9	9	CTRL I	01001	HT	Horizontal Tab
10	A	CTRL J	01010	LF	Line Feed
11	B	CTRL K	01011	VT	Vertical Tab
12	C	CTRL L	01100	FF	Form Feed
13	D	CTRL M	01101	CR	Carriage Return
14	E	CTRL N	01110	SO	Shift Out
15	F	CTRL O	01111	SI	Shift In
16	10	CTRL P	10000	DLE	Date Link Escape
17	11	CTRL Q	10001	DC1	Device Control 1
18	12	CTRL R	10010	DC2	Device Control 2
19	13	CTRL S	10011	DC3	Device Control 3
20	14	CTRL T	10100	DC4	Device Control 4
21	15	CTRL U	10101	NAK	Negative Acknowledge
22	16	CTRL V	10110	SYN	Synchronous Idle
23	17	CTRL W	10111	ETB	End of Transmission Block
24	18	CTRL X	11000	CAN	Cancel
25	19	CTRL Y	11001	EM	End Medium
26	1A	CTRL Z	11010	SUB	Substitute or EOF End Of File
27	1B		11011	ESC	Escape
28	1C		11100	FS	File Separator
29	1D		11101	GS	Group Separator
30	1E		11110	RS	Record Separator
31	1F		11111	US	Unit Separator

ASCII text characters

This table was produced using Excel 97 running on Microsoft Windows operating system. ASCII values 128 to 255 are not standard so other operating systems may yield different results. ASCII is based on the first 7 bits so affect values 0–127.

Dec	Hex	Binary	ASCII	Dec	Hex	Binary	ASCII	Dec	Hex	Binary	ASCII
32	20	100000		107	6B	1101011	k	182	B6	10110110	
33	21	100001	!	108	6C	1101100	l	183	B7	10110111	·¶
34	22	100010	"	109	6D	1101101	m	184	B8	10111000	¸
35	23	100011	#	110	6E	1101110	n	185	B9	10111001	¹
36	24	100100	$	111	6F	1101111	o	186	BA	10111010	º
37	25	100101	%	112	70	1110000	p	187	BB	10111011	»
38	26	100110	&	113	71	1110001	q	188	BC	10111100	¼
39	27	100111	'	114	72	1110010	r	189	BD	10111101	½
40	28	101000	(	115	73	1110011	s	190	BE	10111110	¾
41	29	101001	)	116	74	1110100	t	191	BF	10111111	¿
42	2A	101010	*	117	75	1110101	u	192	C0	11000000	À
43	2B	101011	+	118	76	1110110	v	193	C1	11000001	Á
44	2C	101100	,	119	77	1110111	w	194	C2	11000010	Â
45	2D	101101	-	120	78	1111000	x	195	C3	11000011	Ã
46	2E	101110	.	121	79	1111001	y	196	C4	11000100	Ä
47	2F	101111	/	122	7A	1111010	z	197	C5	11000101	Å
48	30	110000	0	123	7B	1111011	{	198	C6	11000110	Æ
49	31	110001	1	124	7C	1111100	|	199	C7	11000111	Ç
50	32	110010	2	125	7D	1111101	}	200	C8	11001000	È
51	33	110011	3	126	7E	1111110	~	201	C9	11001001	É
52	34	110100	4	127	7F	1111111	□	202	CA	11001010	Ê
53	35	110101	5	128	80	10000000	€	203	CB	11001011	Ë
54	36	110110	6	129	81	10000001	□	204	CC	11001100	Ì
55	37	110111	7	130	82	10000010	‚	205	CD	11001101	Í
56	38	111000	8	131	83	10000011	ƒ	206	CE	11001110	Î
57	39	111001	9	132	84	10000100	„	207	CF	11001111	Ï
58	3A	111010	:	133	85	10000101	…	208	D0	11010000	Ð
59	3B	111011	;	134	86	10000110	†	209	D1	11010001	Ñ
60	3C	111100	<	135	87	10000111	‡	210	D2	11010010	Ò
61	3D	111101	=	136	88	10001000	^	211	D3	11010011	Ó
62	3E	111110	>	137	89	10001001	‰	212	D4	11010100	Ô
63	3F	111111	?	138	8A	10001010	Š	213	D5	11010101	Õ
64	40	1000000	@	139	8B	10001011	‹	214	D6	11010110	Ö
65	41	1000001	A	140	8C	10001100	Œ	215	D7	11010111	×
66	42	1000010	B	141	8D	10001101	□	216	D8	11011000	Ø
67	43	1000011	C	142	8E	10001110	Ž	217	D9	11011001	Ù
68	44	1000100	D	143	8F	10001111	□	218	DA	11011010	Ú
69	45	1000101	E	144	90	10010000	□	219	DB	11011011	Û
70	46	1000110	F	145	91	10010001	'	220	DC	11011100	Ü
71	47	1000111	G	146	92	10010010	'	221	DD	11011101	Ý
72	48	1001000	H	147	93	10010011	"	222	DE	11011110	Þ
73	49	1001001	I	148	94	10010100	"	223	DF	11011111	ß
74	4A	1001010	J	149	95	10010101	•	224	E0	11100000	à
75	4B	1001011	K	150	96	10010110	–	225	E1	11100001	á
76	4C	1001100	L	151	97	10010111	—	226	E2	11100010	â
77	4D	1001101	M	152	98	10011000	~	227	E3	11100011	ã
78	4E	1001110	N	153	99	10011001	™	228	E4	11100100	ä
79	4F	1001111	O	154	9A	10011010	š	229	E5	11100101	å
80	50	1010000	P	155	9B	10011011	›	230	E6	11100110	æ
81	51	1010001	Q	156	9C	10011100	œ	231	E7	11100111	ç
82	52	1010010	R	157	9D	10011101	□	232	E8	11101000	è
83	53	1010011	S	158	9E	10011110	ž	233	E9	11101001	é
84	54	1010100	T	159	9F	10011111	Ÿ	234	EA	11101010	ê
85	55	1010101	U	160	A0	10100000		235	EB	11101011	ë
86	56	1010110	V	161	A1	10100001	¡	236	EC	11101100	ì
87	57	1010111	W	162	A2	10100010	¢	237	ED	11101101	í
88	58	1011000	X	163	A3	10100011	£	238	EE	11101110	î
89	59	1011001	Y	164	A4	10100100	¤	239	EF	11101111	ï
90	5A	1011010	Z	165	A5	10100101	¥	240	F0	11110000	ð
91	5B	1011011	[	166	A6	10100110	¦	241	F1	11110001	ñ
92	5C	1011100	\	167	A7	10100111	§	242	F2	11110010	ò
93	5D	1011101	]	168	A8	10101000	¨	243	F3	11110011	ó
94	5E	1011110	^	169	A9	10101001	©	244	F4	11110100	ô
95	5F	1011111	_	170	AA	10101010	ª	245	F5	11110101	õ
96	60	1100000	`	171	AB	10101011	«	246	F6	11110110	ö
97	61	1100001	a	172	AC	10101100	¬	247	F7	11110111	÷
98	62	1100010	b	173	AD	10101101	-	248	F8	11111000	ø
99	63	1100011	c	174	AE	10101110	®	249	F9	11111001	ù
100	64	1100100	d	175	AF	10101111	¯	250	FA	11111010	ú
101	65	1100101	e	176	B0	10110000	°	251	FB	11111011	û
102	66	1100110	f	177	B1	10110001	±	252	FC	11111100	ü
103	67	1100111	g	178	B2	10110010	²	253	FD	11111101	ý
104	68	1101000	h	179	B3	10110011	³	254	FE	11111110	þ
105	69	1101001	i	180	B4	10110100	´	255	FF	11111111	ÿ
106	6A	1101010	j	181	B5	10110101	µ				

Appendix B
Fact sheet

How to obtain a hex dump with DEBUG

Debug is a very odd program! It can be used to look into RAM directly or into any file, it can assemble or unassemble files, it can run programs and can be used to change as little as 1 bit.

It is started at the DOS prompt by

C:\>DEBUG yourfile.txt

and all you get is a '–' character!

This '–' character is the input prompt. Possible commands are shown below.

```
assemble          A [address]
compare           C range address
dump              D [range]
enter             E address [list]
fill              F range list
go                G [=address] [addresses]
hex               H value1 value2
input             I port
load              L [address] [drive] [firstsector] [number]
move              M range address
name              N [pathname] [arglist]
output            O port byte
proceed           P [=address] [number]
quit              Q
register          R [register]
search            S range list
trace             T [=address] [value]
unassemble        U [range]
write             W [address] [drive] [firstsector] [number]
allocate expanded memory        XA [#pages]
deallocate expanded memory      XD [handle]
map expanded memory pages       XM [Lpage] [Ppage] [handle]
display expanded memory status  XS
```

For the use that DEBUG is put to here, to produce a hex dump, the only commands needed are D for Dump and Q for Quit.

To produce a hex dump of a file called TEST.DAT simply issue the command

C:\> DEBUG TEST.DAT

at the command prompt then use the command D (then enter). You can 'dump' from any address and any size, the command D 100 1000 dumps from address 100 and dumps 1000 bytes, both numbers in hex.

If you need the hex dump in a file, you should first prepare a command file that contains just the data below. Assume this command file is called CMD.DAT, it is made using the commands at the DOS prompt

C:\> COPY CON CMD.DAT
D
Q
CTRL Z

where CTRL Z means press the CTRL and Z keys together.

You can now produce a hex dump in a file by using I/O redirection as in the command:

C:\> DEBUG TEST.DAT < CMD.DAT > TESTHEXDUMP.TXT

The < character means take input from your command file called CMD.DAT and the > character means place or redirect the output into whatever file name you provide, in this case TESTHEXDUMP.TXT.
   If you need to dump more than 100 hex bytes, the file CMD.DAT will have to contain

```
D 100 200
Q
CTRL Z
```

where the 200 refers to 200 hex bytes in length. If you need to quote a length in the D command, the starting address (usually 100) must also appear.
   See:

http://www.geocities.com/thestarman3/asm/debug/debug.htm
or
http://www.ping.be/~ping0751/debug.htm

for more information on DEBUG.